Annual Conference of the American Forage and Grassland Council (AFGC 2019)

Forages Yesterday, Today, and Tomorrow

St. Louis, Missouri, USA
6-9 January 2019

ISBN: 978-1-5108-9275-0

Printed from e-media with permission by:

Curran Associates, Inc.
57 Morehouse Lane
Red Hook, NY 12571

Some format issues inherent in the e-media version may also appear in this print version.

Printed by Curran Associates, Inc. (2019)

For permission requests, please contact American Forage and Grassland Council (AFGC) at the address below.

American Forage and Grassland Council (AFGC)
PO Box 867
Berea, KY 40403
USA

Phone: 800.944.2342

www.afgc.org

Additional copies of this publication are available from:

Curran Associates, Inc.
57 Morehouse Lane
Red Hook, NY 12571 USA
Phone: 845-758-0400
Fax: 845-758-2633
Email: curran@proceedings.com
Web: www.proceedings.com

TABLE OF CONTENTS

AGRONOMIC AND NUTRITIVE VALUE OF COOL-SEASON ANNUAL MIXTURES FOR BALEAGE PRODUCTION

S. L. Shoup, R.B. Muntifering, M.K. Mullenix, S.L. Dillard[1]

Baleage is a high moisture feed that is gaining popularity in the Southeast due to its consistent nutrient quality as well as its ability to be harvested at optimum maturity. The objective of this study was to compare the agronomic and nutritive value of two cool-season annual mixtures, wheat (*Triticum aestivum*) and T-Raptor (*Brassica rapa* × *napus*; T), or wheat and crimson clover (*Trifolium incarnatum*; C) with (I) or without (N) inoculant. The baleage was tested at 8 time periods after ensiling. There was no difference in ADF concentration among treatments, but day 120 after ensiling was greater ($P \leq 0.0024$) than all other sampling dates. There was no difference in NDF between forages and silage inoculant treatments or among days. Crude protein of C was greater ($P \leq 0.0001$) than T by 3.6 units. The pH of C was greater ($P \leq 0.001$) than T, but not different between I and N treatments. The pH did increase at day 120 due to the lack of long-term anaerobic stability but was within normal range at day 28. Lactic acid concentration interactions for forage and silage, and silage by day were not different ($P = 0.0372$, $P \leq 0.0001$ respectively). There was no difference ($P = 0.2633$) between forage treatments for acetic acid however, forage treatments differed ($P \leq 0.0001$) with I being greater than N (5.49 vs. 1.94 %, respectively). Acetic acid increased ($P \leq 0.0001$) across all days. Butyric acid concentrations were greater ($P \leq 0.056$) for C as a result of the buffering capacity as well as pH differences. These results suggest that silage inoculants in these mixtures are economically viable, but that these mixtures are suitable for baleage production and provide adequate forage quality to maintain animal performance.

[1]Dept. of Animal Sciences, Auburn University, Auburn, AL (sls0017@auburn.edu).

ALTERNATIVE WARM-SEASON FORAGE SYSTEMS FOR WEANED BEEF HEIFERS IN THE ALABAMA BLACK BELT SOIL REGION

M.L. Marks[1], C.G. Nichols, M.K. Mullenix, J. Yeager, K.B. Mason and R.B. Muntifering
Auburn University/Alabama Cooperative Extension System, Auburn, AL

Endophyte-infected tall fescue (ETF) is the predominant perennial forage used by beef cattle producers in the Black Belt prairie soil region; therefore most beef heifers are developed using ETF as the forage source during the development phase of their production. Summer perennial grasses could provide an alternative forage option to reduce the impacts of ETF on heifer performance. The objective of this study was to evaluate herbage mass and animal performance of weaned beef heifers grazing a mixture of big bluestem, little bluestem, and indiangrass (NWSG), ETF, or 'Pensacola' bahiagrass (PB) at the Black Belt Research and Extension Center (BBREC) in Marion Junction AL. Paddocks were continuously stocked with four weaned Angus × Simmental beef heifers from early June through late August 2018 (71 days of grazing). Additional put-and-take heifers were used to manage forage to a target herbage mass of 2,000 lb DM/acre. Forage mass was collected every two weeks during the trial and animal weights were collected every 28-d. Year one data indicates PB (2,812 DM lbs/acre) and ETF (2,168 DM lbs/acre) had greater yields compared to the NWSG system (1,777 DM lbs/acre). No differences were observed among the systems for initial BW, final weight, ADG, BCS across the grazing season illustrating similarity across forage systems. Year one data indicates summer perennial grasses could be used as an alternative forage system for developing beef heifers, but supplementation may be needed to sustain desirable levels of gain. Heifers from this project will be tracked through breeding to evaluate conception rates and longevity in the herd.

[1]M. Landon Marks, Regional Extension Agent, Alabama Cooperative Extension System, Centre, AL (mlm0013@auburn.edu).

AN EVALUATION OF THE EFFECTS OF ANNUAL RYEGRASS CYTOTYPE ON HERBAGE MASS, NUTRITIVE VALUE, AND CATTLE PERFORMANCE

S. M. Montgomery[1], J. I. Morrison[1], B. B. Karisch[2], B. S. Baldwin[1], and J. A. Parish[3]

Annual ryegrass (*Lolium multiflorum* Lam.) can be divided into two cytotypes: diploid (2n = 2x) or tetraploid (2n = 4x). Improving genetics of diploid to tetraploids results in increased seedling vigor, more robust growth and thus supposedly greater yield. There has been conflicting research that suggests no difference in yield among diploid and tetraploid cultivars. If there is no difference in yield between cytotypes, then is there an advantage in weight gain of cattle grazing tetraploid cytotype? To determine if there is an advantage in weight gain of cattle grazing diploid and tetraploid annual ryegrass, this research project took place at the H. H. Leveck Animal Research Center in Starkville, MS. Two diploid annual ryegrass cultivars ('Marshall' and 'Tam 90') and two tetraploid cultivars ('Jumbo' and 'Nelson') were planted in 2 acre pastures replicated four times. Four-steers were randomly assigned to each of the 16 pastures for an 82-d grazing period. Initial and end body weights were taken over a two-day period then averaged for each steer. Weights were taken starting at day 0 then every 28-d throughout the season. Final ADG for cattle grazing 'Tam 90' (2.43 lbs/d) were greater than 'Marshall' (2.03 lbs/d). Variability around the mean for cattle grazing 'Nelson' (2.29 lbs/d) and 'Jumbo' (2.18 lbs/d) fell into the same category as 'Tam 90' and 'Marshall'. Other parameters measured included: visual emergence ratings, dry matter yield, and nutritive value.

[1]Masters Graduate Student, Dept. of Plant & Soil Sciences, Mississippi State University, MS (smm1006@msstate.edu); [1]Assistant Research Professor, Dept. of Plant & Soil Sciences, Mississippi State University, MS,(jim46@pss.msstate.edu); [2]Associate Extension/ Research Professor, Dept. of Animal and Dairy Sciences, Mississippi State University, MS, (brand.karish@msstate.edu); [1]Professor, Dept. of Plant & Soil Sciences, Mississippi State University, MS,(bsb2@pss.msstate.edu). [3]Professor and Head, North Mississippi Research and Extension Center, Verona, MS, (j.parish@msstate.edu)

AN INTEGRATED CROP-LIVESTOCK SYSTEM FOR EASTERN NEBRASKA

E. Widder[1], R. Mitchell[2], D. Redfearn[1], V. Jin[3], M. Schmer[3]

In eastern Nebraska, producers could integrate livestock and perennial pasture into grain cropping systems to potentially increase revenue. Our objective was to develop an integrated crop-livestock production system for eastern Nebraska on marginally productive land. In 2014, a 50-acre field-scale demonstration site was established, including 10-acre pastures of 'Newell' smooth bromegrass, 'Shawnee' switchgrass, and 'Liberty' switchgrass and 20 acres of corn near Mead, NE. Forage was harvested in 2016, 2017, and 2018. Forage yield for Newell in 2016 was 2.2 tons/acre; forage yield for Liberty switchgrass was 3.0 and 3.6 tons/acre in 2016 and 2017; Shawnee switchgrass forage yield was 2.4 and 0.9 tons/acre in 2016 and 2017; corn stover yield was 0.7 and 1.0 tons/acre in 2016 and 2017. Pastures were grazed by 18 yearling steers. In 2017, the grazing rotation was Newell (May 18-June 8) with an average daily gain/head/day (ADG) of 2.4 lb/day, Liberty (June 13-20), Shawnee (June 20-Sept. 1) with an ADG of 0.9 lb/day, and Newell (Sept. 7-Oct. 6). Triticale cover crop was planted after corn residue harvest in 2016 and 2017, but grazable forage was not produced. In 2018, the grazing rotation began with Newell (May 2-June 1) with an ADG of 1.2 lb/day. The herd was split to graze Shawnee (9 steers) and Liberty (9 steers) (June 11-August 29), with an ADG of 0.9 and 1.3 lb/day, respectively. The herd was combined and returned to Newell (Sept.5-18) with an ADG of 0.4 lb/day. Corn yields in 2016, 2017, and 2018 were 120, 155, and 131 bu/acre respectively. The study is continuing and demonstrates a climate resilient crop-livestock production system for eastern Nebraska.

[1]Department of Agronomy and Horticulture, University of Nebraska-Lincoln (UNL), Lincoln, NE [2]USDA-ARS Wheat, Sorghum, and Forage Research Unit, UNL, Lincoln, NE [3]USDA-ARS Agroecosystem Management Research Unit, UNL, Lincoln, NE.

ANALYSIS OF SOIL HEALTH OF PERENNIAL PASTURES UNDER DIFFERING YEARS OF WINTER ANNUAL VEGETATION AND GRAZING

K. M. Bridges, L. M. Fultz, and M. W. Alison[1]

Utilizing winter annuals in livestock pastures reduces feed costs of hay, and provides additional vegetative cover that is beneficial to the soil. However, how this practice affects the soil microbial communities has not been well studied. This project was conducted on three different pastures that had been under cattle grazing for varying amounts of time: 10 years, 4 years, and 0 years. Additionally, the 10 and 4 year pastures had had winter annuals incorporated for the past 4 years while the ungrazed pasture had no management improvements. Each pasture was divided into three pseudo-replicates from which five soil cores were collected to a depth of 30 cm. Samples were analyzed for soil nutrient content, soil microbial community composition and enzymatic activity, and bulk density. According to years of management, soil pH was highest at 5.95, bulk density was 6% lower, and β-glucosidase 25% higher in the 10 year pasture compared to the other pastures. According to date of collection, total carbon decreased by 40% from fall 2016 to fall 2017 before recovering in spring 2018 while total nitrogen remained unchanged before increasing by 31% in spring 2018. The fungi to bacteria ratio (F:B) decreased fivefold from fall 2016 to spring 2017. As the soil microbial community structure, soil organic matter, TC, and TN was unchanged across years under management, it may be that an equilibrium was achieved across these systems. Also, the changes in the soil microbial community by date of collection may be in response to abiotic and biotic factors.

[1]Graduate Student, Louisiana State University, Baton Rouge, LA (kbridges@agcenter.lsu.edu); Assistant Professor, Soil Microbiology, Louisiana State University, Baton Rouge, LA (lfultz@agcenter.lsu.edu); Extension Specialist, LSU Scott Research and Extension Center, Winnsboro, LA (walison@agcenter.lsu.edu)

CAN TARGETED MANAGEMENT REDUCE NIMBLEWILL IN PASTURES?

K.L. Lea and S. Ray Smith[1]

Nimblewill, *Muhlenbergia schreberi*, is a native, warm season, perennial grass in central Kentucky known for its aggressive growth in heavily grazed pastures. Livestock, including horses, cattle and goats, are not known to consume nimblewill, leaving it to persist and spread in pastures unchecked. Currently, there are no herbicides labeled for pasture use in Kentucky to control this grass in a grass stand. The goal of this project was to evaluate the possibility of using competition and shade to reduce nimblewill presence in horse pastures. Four treatments (tall fescue, perennial ryegrass, white clover and a control) were seeded into thick stands of nimblewill in the fall of 2017 in a randomized complete block experimental design. Throughout the 2018 growing season, the plots were split and mowed at three different heights: Low (~3 in), high (~8 in) and hay (only harvested twice in the season). Plots were sampled each month for canopy height and species composition by visual estimation. Three, 1-ft^2 samples were taken from each plot in August of 2017 and 2018; samples were hand separated by species and weighted to determine species composition. Preliminary analysis of the data suggests that nimblewill presence was reduced in subplots that were mowed high or managed for hay, suggesting shade from other grasses can affect nimblewill growth. Overseeded species had no effect on nimblewill growth. Further research is needed to confirm these preliminary results and to characterize the amount of shade that may be needed to reduce nimblewill in pastures.

[1] Research Analyst(Krista.lea1@uky.edu)and Extension Professor (raysmith1@uky.edu)
University of Kentucky, Dept. of Plant and Soil Sciences, Lexington, KY).

CARBOHYDRATE RECOVERY IN GRASS SPECIES UNDER TWO DEFOLIATION FREQUENCIES DETERMINED BY THERMAL TIME

I. Calvache[1], O.A. Balocchi[2], M.F. Alonso[2], and I.F. López[3]

Plant recovery is a dynamic process that varies throughout the year and is affected by pasture defoliation frequency. In grass species, plant water soluble carbohydrates (WSC) reserves recovery after defoliation and plant ability to mobilize WSC have been little studied. The objective was to evaluate the fluctuation of the WSC concentration in *Lolium perenne* L. (Lp) and *Bromus valdivianus* Phil. (Bv) subjected to two defoliation frequencies determined by accumulated growing degree-days (AGDD). The study was conducted at the Universidad Austral de Chile's Austral Research Station in autumn 2017, on Lp and Bv pastures. Accumulated herbage mass (kg DM/ha), WSC concentration in stubble and leaves (g/1000g) were measured every 3 days. The high defoliation frequency (135 AGDD) treatment initially replenished CHOS concentration levels faster than the low defoliation frequency (270 AGDD) treatment. Lp leaves had a higher concentration of WSC than Bv leaves when defoliated at 135 AGDD. Accumulated herbage mass was lower for both species when defoliated at 135 AGDD compared to 270 AGDD. Plant recovery and persistence of Lp and Bv had a strong positive relationship with defoliation at 270 AGDD. This study validated AGDD as tool to determine defoliation frequency considering the WSC utilization and recovery and plant growth post-defoliation.

[1]Postgraduate School, Faculty of Agricultural Science, Universidad Austral de Chile, Valdivia, Chile (ivan.calvache@postgrado.uach.cl); [2]Institute of Animal Production, Universidad Austral de Chile, Valdivia, Chile (obalocch@uach.cl; maximo.alonso@uach.cl); [3]School of Agriculture and Environment, Massey University, Palmerston North, New Zealand (I.F.Lopez@massey.ac.nz). Funded by Fondecyt: 1141016.

COVER CROPS: NUTRITIONAL VALUES AND RELATED FEEDING ISSUES

D. R. Oswald[1]

Elevated nutritional values of fresh cover crop forages can greatly increase feeding-related animal health issues when abrupt changes are made in the animal diet. Selected cover crop species were tested to determine forage quality. Cover crop grasses and brassica crops tested had high moisture (79-92%); high crude protein (CP 17-30%); low neutral detergent fiber (NDF 23-39%) and high relative feed value (RFV 188-292) 90 days after seeding. Cereal rye, wheat, triticale, spring and winter oats, spring and winter barley, and annual ryegrass cover crops can cause bloat, grass tetany, nitrate or fertility issues. Thiamine deficiency, nitrate poisoning, bloat and digestive upsets in ruminant animals can be potential problems when feeding brassica crops like rape, turnip tops and bulbs, and radish tops and bulbs. High or excess soil fertility may increase potential health issues. Management is the key to preventing the health issues when feeding high quality fresh forages to ruminants. Forage testing is an essential part of prevention. Dietary changes in forage quality should be made slowly to allow rumen microbes to adjust to feeding practices.

[1]Cover Crop Specialist, Zea Mays Foundation, Forage/Grazing Specialist, Midwest Grass & Forage, Kewanee, IL (oswalddean@gmail.com).

CRABGRASS CULTIVAR RESPOINSE TO NITROGEN FERTILIZER ADDITIVES

R. Lemus[1] and B. Rushing[2]

In the southern USA, crabgrass (*Digitaria sanguinalis*) is a valuable summer annual forage that can be used for grazing or hay production, has high nutritive value, and is adapted to a variety of soil and climatic conditions. The objective of this study was to evaluate cultivar response to nitrogen fertilizer additives. The study was conducted at Mississippi State University on a Marietta Fine Sandy loam soil. Five crabgrass cultivars ('Dal's Big River' (DBR), 'NFC01-1' (NFC), 'Quick-N-Big' (QNB), 'Quick-N-Big Spreader' (QNBS), and 'Red River' (RR) were drilled in a prepared seed bed at 8 lb PLS/ac. Cultivars received four nitrogen combinations: control, urea ammonium nitrate (UAN, 32-0-0), UAN plus N-Veil® (26.7% NBTP), and UAN plus Preserve N™ (18-0-0). Nitrogen was applied at rate of 50 lb N/ac. UAN was treated with N-Veil® and Preserve N™ at a rate of 2.0 qt/ton and applied using a sprayer when plants reached a 3-in height to supply 25 gal/ac. Lime, P, and K were applied based on soil test recommendations. Plots were harvested once in 2018 when plants reached 12-in height. Subsamples were dried and analyzed for nutritive value using NIRS. Biomass production was affected by cultivar (P = 0.0201) where QNBS (highest yielding) had 24% greater biomass production than NFC (lowest yielding). There was no N treatment effect on biomass production, but treatments containing fertilizer additives were lower yielding than the control. No significant differences were noted for cultivar and treatment effects on CP and ADF. Species were significantly different in NDF concentration (P = 0.0074).

[1]Extension Forage Specialist, Center for Forage Management & Environmental Stewardship, Mississippi State University (Rocky.Lemus@msstate.edu); Assistant Research/Extension Professor, Coastal Plain Experiment Station, Mississippi State University (Brushing@pss.msstate).

DETECTING AND MONITORING BERMUDAGRASS STEM MAGGOT POPULATIONS

Lisa Baxter, Holly Anderson, Andy Carter, Michasia Dowdy, Brooke Hampton, Stephanie Hollifield, Carole Knight, Will Lovett, Kelly Paulk, Pamela Sapp, Andrew Sawyer, and Dennis Hancock[1]

The bermudagrass stem maggot (BSM; *Atherigona reversura* Villenueve) continues to be a major pest problem throughout the Southeast US. Although strategic pyrethroid insecticide applications are an effective control strategy, much work is needed to fine-tune these recommendations. County Extension Agents and Coordinators from eight counties across South GA participated in an on-farm trial to determine the best method for monitoring BSM activity. This trial tested the use of sweep nets and sticky card traps during different times of the day (8:00 to 11:00 AM, 11:00 AM to 2:00 PM, and 2:00 to 5:00 PM). The BSM were most active at the first time point (8:00 to 11:00 AM) and a greater number of flies are caught using the sweep net. More BSM flies were observed on the sticky cards at the 8" height than the 16" or 24". Although the objective of this project was to determine the best method for monitoring the BSM, we have gained valuable insight in how to optimize the efficiency of insecticide applications. These data show that insecticide applications would be most effective at or before 11:00 AM and should be sprayed deep into the canopy.

[1]Post-doctorate, Crops and Soils Science Dept., University of Georgia, Tifton, GA (baxterl@uga.edu); Ben Hill County Extension Agent, Fitzgerald, GA, (hhickey@uga.edu); Lowndes County Extension Agent, Valdosta, GA, (aec22711@uga.edu); Brooks County Extension Agent, Quitman, GA, (michasia@uga.edu); Wheeler County Extension Coordinator, Alamo, GA, (brooke@uga.edu); Brooks County Extension Coordinator, Quitman, GA (smh@uga.edu); Bulloch County Extension Agent, Statesboro, GA, (clh@uga.edu); Bacon County Extension Coordinator, Alma, GA, (welovett@uga.edu); Books County Extension Intern, Quitman, GA; Jefferson County Extension Coordinator, Louisville, GA, (pamsapp@uga.edu); Wilcox County Extension Coordinator, Rochelle, GA (agsawyer@uga.edu); Forage Extension Specialist, Crop and Soils Science Dept., University of Georgia, Athens, GA, (dhancock@uga.edu)

DETERMINING THE EFFECTS OF PLANTING DATE AND LAND PREPARATION METHOD ON THE FORAGE YIELD AND QUALITY OF FORAGE BRASSICA

Tayler Denman[1], Dennis Hancock[2], and J. D. Hale[3]

Forage-based livestock producers in the Southeastern US have a competitive advantage in the industry because warm summers and mild winters allow for an extended growing season compared to other parts of the US; however, there is still a gap from October through December, when few forages grow. Forage brassicas are rapidly establishing cool season crops that are used commonly as a forage in New Zealand. It is hypothesized that these characteristics could potentially fill the forage gap in October through November in Georgia where winters are mild however, there is little information on basic agronomic practices such as planting date and land preparation method on the forage yield and quality of forage brassica, especially when grown in Georgia. The purpose of this experiment is to evaluate the effect of land preparation method (Conventional till, No till – physical burning, No till – with mowing, and No till – no residue removal) and planting dates (September 1st, September 15th, October 1st, and October 15th) on the forage yield and quality of forage brassica when grown in Georgia. Results indicated that earlier planting dates (September 1st and September 15th) produced significantly higher forage yields than later planting dates (October 1st and October 15th) in addition, Planting into as little residue as possible (Conventional till and No till – with physical burning) produced significantly higher forage yields than treatments with more residue remaining (No till – with mowing and No till without residue removal). The nutrient quality of brassica was observed to be sufficient for livestock with high nutrient demands.

[1]Graduate Student/Program Coordinator, University of Georgia, Athens, GA; [2]Professor and Forage Extension Specialist, University of Georgia, Athens, GA, [3]Research Technician, University of Georgia, Athens, GA.

DEVELOPING A FORAGE GRASS MATURITY INDEX

Ray Smith, Roeland Kapsenberg, Jerome Magnuson, Steve Reid, Patton Slusher, Jerry Hall, Brent Jones, Robin Lamp, Yedilaklil Hunde, and Devesh Singh[1]

There is a wide range of reproductive maturity among forage grass species and varieties sold in the U.S., but no uniform rating system for selecting varieties. The AFGC Forage Grass Maturity Working Group is comprised of seed industry and University representatives who are developing a forage grass maturity index. To provide initial maturity information, 8 grass species and 72 varieties were planted at company research farms in Oregon (Barenburg USA), Iowa (Grassland Oregon), and Kentucky (DLF). Species included annual, hybrid, and perennial ryegrass, Kentucky bluegrass, bromegrass sp., orchardgrass, tall fescue and timothy; varieties were selected to cover the range of known maturities within each species. Plot design followed an adapted protocol used for Grass Variety Review Board (GVRB)/Plant Variety Protection (PVP) applications, with a minimum of 30 space plants per variety over 3 replications, use of Julian calendar dates and growing degree days, maturity date observations when 3 reproductive tillers visible, and measurements the 2nd summer after seeding perennials. Results in 2017 and 2018 showed that Kentucky bluegrass started maturing on day 112 and varieties of timothy and perennial ryegrass were still maturing on day 161. There was considerable overlap across species, variety and location and a 20-day maturity range across varieties within 4 species. The working group proposes that company/University breeders apply to AFGC for their variety to be added to the AFGC-Maturity Index list after providing proof of variety acceptance by GVRB/PVP and providing the maturity dataset collected per the established protocol.

[1]Ray Smith, Univ. of KY (raysmith1@uky.edu); Roeland Kapsenberg, Jerome Magnuson, Steve Reid, and Patton Slusher, DLF Pickseed; Jerry Hall and Brent Jones, Grassland Oregon; Robin Lamp and Yedilaklil Hunde, Barenbrug USA; and Devesh Singh, Wilbur-Ellis.

DEVELOPMENT OF GRAZING RECOMMENDATIONS FOR MANAGING ALFALFA-BERMUDAGRASS MIXTURES IN THE SOUTHEASTERN U.S.

K.R. Groce, J.J. Tucker, K. B. Mason, L. Marks, and M.K. Mullenix[1]

There has been growing interest in interseeding high-quality legumes, like alfalfa, into existing bermudagrass stands as a step towards improving forage, animal, and ecosystem stability in the Southeast. Much of the work evaluating these mixtures has been focused on management and production of stored forages (i.e. dry hay/baleage), and limited work has evaluated alfalfa-bermudagrass mixtures under grazing. It has been noted that grazing intensity and frequency impact overall stand production over time, but no published work has evaluated the relationship between harvest height and frequency to better define grazing recommendations. The objective of this research is to evaluate the effect of harvest height and frequency of newly established alfalfa interseeded into an existing hybrid bermudagrass stand, and begin to develop grazing recommendations for management of these mixtures. In 2018, thirty-six alfalfa-bermudagrass plots were established at Crossville and Shorter, AL and Watkinsville and Tifton, GA using a randomized complete block design with four replications at each location. Alfalfa varieties were selected for suitability to location and were either 'Bulldog 505' or 'Bulldog 805'. Treatments included combinations of harvest frequency (2, 4, or 6 weeks) and harvest height (2, 4, or 6 inches), and plots were evaluated for their influence on sward yield, persistence, stand density, botanical composition, nutritive value, and change over time. It is expected that the 'Bulldog 805' 6-week, 2-inch plots will have the greatest average yield (lb/ac) over the season and the 'Bulldog 505' 2-week, 6-inch plots will have the lowest average yield (lb/ac) over the season. This study is supported by funding from the USDA-NIFA-Alfalfa Forage Research Program (2017).

[1]Graduate Research Assistant, Univ. of Georgia, Dept. Animal and Dairy Science, Tifton, GA (kaylyn.groce@uga.edu); Assistant Professor, Univ. of Georgia, Dept. Animal and Dairy Science, Tifton, GA (jjtucker@uga.edu); Graduate Research Assistant, Auburn Univ., Dept. Animal Sciences, Auburn, AL (kmb0108@auburn.edu); Regional Ext. Agent, Auburn Univ., Centre, AL (mlm0013@aces.edu); Assistant Professor, Auburn Univ., Dept. Animal Sciences, Auburn, AL (mullemk@auburn.edu).

DIAGNOSIS AND RESPONSE TO FESCUE TOXICOSIS IN A CENTRAL KENTUCKY HORSE FARM: AN EXTENSION CASE STUDY.

Jimmy C. Henning, Krista Lea, Ray Smith and Chris Teutsch[1]

A central Kentucky thoroughbred horse farm experienced a high incidence of fescue toxicosis symptoms in pregnant and foaling mares during early 2017. The University of Kentucky Plant and Soil Science Forage Group was contacted by the farm at the advice of the consulting veterinarian. After a farm visit, the farm agreed to a program of pasture evaluation. Fields were sampled for species composition, endophyte infection level, and ergovaline content. Toxic levels of tall fescue were found and the farm began an aggressive program of fescue eradication in selected fields to eliminate fescue toxicosis. As a result of the work of the UK Horse Pasture Evaluation Program, the farm avoided exposing pregnant mares to toxic tall fescue. As a result, the farm experienced no dystocia (difficulty in foaling), fewer thickened placentas (red bags) and no foal losses due to tall fescue in 2018. Dystocia, thickened placentas and foal loss are indicators of fescue toxicity in pregnant mares. Improved pasture management and avoidance of tall fescue resulted in four more live foals and greatly reduced veterinary costs in 2018 compared to 2017. The economic impact of having four more foals was estimated to be $428,000 using the 2015-17 three-year average Keeneland September yearling sale figures.

[1]Extension Professor, University of Kentucky, Lexington, KY (jimmy.henning@uky.edu); UK Horse Pasture Evaluation Program Coordinator, University of Kentucky, Lexington, KY (krista.lea1@uky.edu); Extension Professor, University of Kentucky, Lexington, KY (raysmith1@uky.edu); Extension Associate Professor, University of Kentucky Research and Education Center, Princeton, KY (chris.teutsch@uky.edu).

DOUBLE CROPPING SUMMER AND FALL SILAGE CROPS FOR BEEF CATTLE

B.L. Powell, K.J. Simon, J.A. Jennings[1]

Double cropping corn grown for silage along with a combination of winter annuals has the potential to be more profitable than corn grown alone as well as beneficial to soil health. Five acres of a combination of Jerry Oats, Elbon Rye and Crimson Clover were drilled September 1st into a clean-tilled seedbed at the rate of 65 lbs/acre of Oats, 50 lb/acre of Cereal Rye and 15 lb/acre of Crimson Clover. Harvesting of the winter annual mix began October 31st using a self-propelled silage cutter. The first harvest consisted mainly of oats with some cereal rye. Once cut, the oats and rye were transported a short distance to a TMR mixer where it was mixed with 2 parts corn silage and 10% Dried Distillers Grain. The resulting ration was then bunk fed to 5 and 6 weight steers to produce a 3 lb/day gain. Temperatures dipped into the upper teens on January 18th and remained below freezing for three days in a row which killed the oats, but the cereal rye was there to take its place so there was no gap in forage production. We saw little forage production out of the Crimson Clover until late spring, but its' added value to soil health made it worth adding to the mix. Due to spring rains, the field flooded in early March; only weeks before planting time. Because of the cover crop, we were able to drill corn into the cover crop as soon as the water was off the field. After the final harvest of the rye, the cover crop was terminated using 1 quart/acre of 41% glyphosate. Forage yields for the cover crop were estimated at 4000 lb/acre of oats and 4000 lb/acre of cereal rye with a combined crude protein of 19.8% and a TDN value of 64.8%.

[1]County Extension Agent – Agriculture, University of Arkansas, Cooperative Extension Service, Danville AR (blpowell@uaex.edu); Animal Science Program Associate – Forages, UACES, Little Rock AR (ksimon@uaex.edu); Animal Science Professor and Extension Forage Specialist, UACES, Little Rock AR (jjennings@uaex.edu).

EFFECTS OF INTERSEEDED CLOVER OR NITROGEN FERTILIZER ON FORAGE METRICS OF ANNUAL RYEGRASS GRAZED BY STOCKER CATTLE

P.A. Gunter, M.K. Mullenix, L.C. Burdette, and R.B. Muntifering[1]

A 3-yr study was conducted to evaluate forage nutritive value and clover abundance when N fertilizer was replaced in part with either interseeded clovers or protein supplements for stocker cattle grazing annual ryegrass (*Lolium* multiflorum). Nitrogen-delivery methods ± monensin included: annual ryegrass fertilized with 112 kg N/ha in a split-application (NFERT), annual ryegrass interseeded with crimson clover and fertilized with 56 kg N/ha at time of establishment (CC), annual ryegrass interseeded with arrowleaf clover and fertilized with 56 kg N/ha at time of establishment (AC), annual ryegrass fertilized with 56 kg N/ha and cattle supplemented with distillers dried grains plus solubles at the rate of 0.65% BW daily (DDGS), and annual ryegrass fertilized at 56 kg N/ha and cattle supplemented with whole cottonseed at the rate of 0.65% BW daily (WCS). Year impacted both *in vitro* true digestibility (IVTD) and crude protein (CP) where IVTD and CP were greatest in Yr 3, intermediate in Yr 2, and least in Yr 1. Degradable intake protein (DIP) was affected by N-delivery method and year where CC was less than NFERT, DDGS, WCS and AC, and Yr 1 was greatest, Yr 2 intermediate, and Yr 3 least. Clover presence was greater for CC than AC across the 3 years of the study, and CC had greater abundance when monensin was fed. Results are interpreted to mean that year had a greater impact on forage nutritive value than N-delivery method.

[1]Phillip A. Gunter, Mary K. Mullenix, and Russell B., Muntifering Upchurch Hall Auburn University, AL 36849, Lance C. Burdette 4725 Co Road Shorter, AL 36075

EFFECTS OF PLANT GROWTH-PROMOTING RHIZOBACTERIA AND NITROGEN FERTILIZER ON STOCKPILED BERMUDAGRASS: YEAR 1

M.E. Griffin[1], S.L. Dillard[1], R.B. Muntifering[1], M.K. Mullenix[1], D.W. Held[2]

Plant growth-promoting rhizobacteria (PGPR) are non-pathogenic, soil inhabiting, beneficial bacteria that colonize the roots of plants. They are widely adapted to different environments and have been shown to increase root and top growth, drought tolerance, and mitigate against pests in turf-type bermudagrass. The purpose of this study was to evaluate two strains of PGPR (Blend 20 and DH44) as an alternative form of N fertilization for fall-stockpiled bermudagrass. In mid-August 2017, eighteen 10-ft^2 plots were mowed to a 1-inch stubble height prior to stockpiling. Treatments included: control, fertilizer, DH44, DH44 + fertilizer, Blend 20, and Blend 20 + fertilizer. Plots were clipped to a height of 1 inch in mid-November, December, and January to determine yield and nutritive value. Forage dry matter (DM) yield for Blend 20 + fertilizer, DH44, and Blend 20 (620, 600, and 593 lb DM/ac, respectively) were greater than the control (534 lb DM/ac). Blend 20 + fertilizer, control, and fertilizer treatments had the greatest effect on crude protein concentration (9.1, 9.5, and 10.1 %, respectively). All treatments had neutral detergent fiber concentrations between 75% and 77%, and acid detergent fiber concentrations between 35.5% and 37.5% with the exception of the control that had less of both fiber fractions. Percentage in vitro true digestibility decreased with the later harvests (46.1, 33.8, and 39.0%, for November, December, and January respectively). The results showed that PGPR-treated plots had increased DM yield and continued to produce more forage biomass through the winter without effecting the overall nutritive quality.

[1]Department of Animal Sciences, Auburn University, Auburn, AL; [2]Department of Entomology and Plant Pathology, Auburn University, Auburn, AL.

ENHANCED EFFICIENCY N FORMULATION EFFECT ON STOCKPILED TALL FESCUE

K.M. Payne, S.R. Smith, and B.M. Goff[1]

Late summer application of nitrogen (N) fertilizer on stockpiled tall fescue has shown to improve forage yields. The use of enhanced efficiency (EE) N fertilizers may be of benefit by reducing ammonia volatilization losses and promoting forage growth later in the growing season. This study evaluated the effect of different EE N formulations (Agrotain treated urea, SuperU, and Environmentally Smart Nitrogen) and untreated urea on the yield and nutritive value of stockpiled 'KY 31' tall fescue over two stockpiling periods (2015-2016 and 2016-2017). Nitrogen was applied at four rates (0, 40, 80, and 120 lb N/acre) in one application in late August to begin the stockpiling period in Lexington, KY. During the 2015-2016 stockpiling period, forage yields increased with increasing N rates; forage yields ranged from 2,600 lb DM/acre when no N was applied and up to 3,689 lb DM/acre when 120 lb N/acre was applied. There was a lower response during the second year, but forage yields increased 75% with the application of N. Forage nutritive value was improved with increasing N rates. There were no differences among EE N sources and standard urea. In summary, N applied in August improved stockpiled forage yield and nutritive value regardless of N source.

[1]Graduate Research Assistant, University of Kentucky, Lexington, KY (kathrynpayne@uky.edu); Extension Forage Specialist, University of Kentucky, Lexington, KY (raysmith1@uky.edu); Agricultural and Natural Resources Extension Agent, West Virginia University, Point Pleasant, West Virginia (BGoff3@mail.wvu.edu).

Establishing A Forage Evaluation Contest for Youth

R. A. Lane[1]

Friendly competition can be a great motivator toward achieving excellence in a given discipline. Since the early 1930s, FFA Career Development Events (CDEs) have built upon what is learned in agricultural classes and encouraged members to put their knowledge into practice. These events are designed to help a member prepare for a career in agriculture by testing and challenging the student's technical, leadership, interpersonal and teamwork skills as well as their knowledge of the subject matter. CDEs answer the question, "When and how will I use this knowledge in the real world?"

Due to necessary commitment of resources and personnel, currently, several institutions host the various career development events at the state level in Texas, listed below:

- Houston Livestock Show & Rodeo - Tractor Tech
- Stephen F. Austin Experimental Forest (NRCS) - Forestry
- Texas Tech University - Ag. Communications, Agronomy, Cotton, Entomology, Food Sci, Horse, Plant ID & Wool
- Region III - State Wildlife CDE
- Tarleton State University - Dairy Cattle, ENR, Homesite, Land, Milk Quality & Rangeland
- Sam Houston State University - Ag. Mech, Ag. Sales, FBM, Floriculture, Forage Evaluation, Marketing Plant & Nursery/Landscape
- Texas A&M University - Livestock, Meats, Poultry & Vet Science

The Texas Forage Evaluation CDE was started as a new contest in 2002. The first year 11 teams competed in the contest at the state level. Twenty-nine teams participated in 2010 and a record 55 teams competed in 2018. It has become quite popular, with over 200 high school student participants each year.

The contest consists of four components: Plant ID (worth 300 points), Hay Judging (300 points), Quiz (300 points), and Problem Solving (100 points). Participants have 25 minutes to complete each component. There are 59 species on the Plant ID list, including a variety of common native and introduced grasses, forage legumes, and weed and brush species. I provide a compact disk with hundreds of photos illustrating identifying features of the various species for a small fee. In the hay judging component, there are four placing classes and 10 hay samples that must be graded for maturity, texture, leafiness, color and foreign matter. The species that may be included in the hay judging component include bermudagrass, bahiagrass, sorghum-sudangrass, alfalfa, mixed grasses, ryegrass, and other winter annual grasses. We developed a handbook on Hay Selection and Evaluation Guidelines, which is provided through the Texas FFA website as a downloadable PDF.

The quiz component includes 30 multiple choice questions derived from the information in two or three chapters of the Southern Forages book by Ball, Hoveland, and Lacefield. The problem-solving component requires mathematical calculations based on practical agricultural situations, such as fencing, fertilization, bale storage, agrochemical applications and others. There are 10 questions on this portion of the contest and students are provided with scratch paper and inexpensive calculators on this section.

While there are national competitions for most of the other FFA career development events, there is not one dealing with forage crops. Youth education in agriculture through FFA and 4-H activities is responsible for the excellent reputation farmers enjoy in this country. Perhaps it is time for AFGC to step in, sponsor and work on the establishment of a forage-based CDE for national competition.

[1]Ph.D., Department of Agricultural Sciences, Sam Houston State University

ESTIMATION OF FORAGE NUTRITIVE VALUES USING UNMANNED AERIAL VEHICLE (UAV)-BASED OPTICAL SENSORS

C. N. Vong[1], J. Zhou[1], J. A. Tooley[2], H. D. Naumann[2] and J. A. Lory[2]

Precision livestock farming is an opportunity to improve forage use efficiency through the timely acquisition of site-specific information. Remote sensing has the potential to estimate forage biomass and nutritive value in real time on pastures. This study evaluated the feasibility of quantifying forage dry matter (DM) and nutritive value using an unmanned aerial vehicle (UAV)-based remote sensing system consisting of a visible and two multispectral cameras. On 28th September 2017, aerial images were taken and plant samples were analyzed for % DM, crude protein (CP), hemicellulose (hemi), acid detergent fiber (ADF), neutral detergent fiber (NDF), NDF digestibility (NDFD), digestible NDF (dNDF), indigestible NDF (iNDF), and in vitro true digestibility (IVTD) using conventional drying and chemical analysis method. Image features such as red (R), green (G), blue (B), hue (H), saturation (S), value (V), red-edge (VIR), normalized difference vegetation index (NDVI), and soil-adjusted vegetation index (SAVI) were extracted as features from the images and used to estimate the DM and nutritive value using statistical models. The results indicated that % DM and CP could be estimated with a coefficient of determination (R^2) of 0.34 and 0.42, respectively. The correlation of image features and other plant quality indicators was low (less than 0.1). There may be potential to improve results using more advanced sensors, such as a hyperspectral camera, which will be tested for future work.

[1]Graduate Student, Division of Food Systems and Bioengineering, University of Missouri (cv6d6@mail.missouri.edu); Assistant Professor, Division of Food Systems and Bioengineering, University of Missouri (zhoujianf@missouri.edu). [2]Graduate Student, Division of Plant Science, University of Missouri (jatc65@mail.missouri.edu); Assistant Professor, Division of Plant Science, University of Missouri (naumannhd@missouri.edu); Professor, Division of Plant Science, University of Missouri (loryj@missouri.edu)

EVALUATING YIELD, NUTRITIVE VALUE, AND STORAGE LENGTH OF TIFTON-85 BERMUDAGRASS AND TIFTON-85 BERMUDAGRASS-ALFALFA MIXTURES AS BALEAGE IN THE SOUTHEAST.

T. J. Hendricks, J. J. Tucker, D. Hancock, R. L. Stewart, Jr.

Interseeding alfalfa (*Medicago sativa*) into bermudagrass (*Cynodon dactylon*) for baleage production can be an effective way to improve forage quality, however quality of baleage stored long-term poses a concern. The objectives of this research are to: 1) compare the yield and nutritive value of bermudagrass with and without interseeded alfalfa, and 2) determine if storage length affects forage quality of baleage. The study utilized an established field of 'Tifton 85' bermudagrass at the University of Georgia Coastal Plains Experiment Station in Tifton, Georgia. Ten 0.5-acre plots were assigned to either T85-only (T85) or T85 interseeded with 'Bulldog 805' alfalfa (T85+Alf). T85 received N fertilization (75 lb N/A) four times throughout the season. Plots were harvested at 10% bloom every 28 to 35 days, baled at 40-60% moisture, and individually wrapped. At harvest, plots were evaluated for stand composition and yield. Bales were sampled prior to wrapping and at 6-weeks, 9- and 12-months post-harvest for nutritive value. Although T85 yields were greater during 2016, additional harvests in T85+Alf plots led to greater ($P < 0.0001$) cumulative yield (32,305 vs. 22,864 lb/A, respectively). Crude Protein and in-vitro digestibility (IVDMD) were greater in T85+Alf treatment than the T85 (14.0 vs 11.4%, $P < 0.001$ for CP; 66.7 vs. 60.2%, $P = 0.03$ for IVDMD, respectively). During storage, CP did not change, while IVDMD of baleage decreased between harvest and 6-weeks.

[1] Graduate Research Assistant, Dept. of Animal and Dairy Science (thend10@uga.edu); Assistant Professor, Dept. of Animal and Diary Science (jjtucker@uga.edu); Associate Professor, Dept. of Crop and Soils (dhancock@uga.edu); Associate Professor, Dept. of Animal and Dairy Science (lawtons@uga.edu), University of Georgia'

EVALUATION OF NITROGEN-DELIVERY METHODS FOR STOCKER CATTLE GRAZING ANNUAL RYEGRASS

P.A. Gunter, M.K. Mullenix, L.C. Burdette, and R.B. Muntifering[1]

A 3-yr grazing experiment was conducted to evaluate efficacy of replacing N fertilizer in part with either interseeded legumes or protein supplementation for stocker cattle grazing annual ryegrass (*Lolium multiflorum*). Each yr, 90 steers (initial BW, 240 ± 13 kg) were assigned to the following N-delivery methods, with or without monensin fed in a free-choice mineral supplement: ryegrass fertilized with 112 kg N/ha (NFERT); ryegrass interseeded with crimson clover (CC, *Trifolium incarnatum*); ryegrass interseeded with arrowleaf clover (AC, *Trifolium vesiculosum*); ryegrass plus distillers dried grains plus solubles (DDGS) supplemented at 0.65% BW daily; and ryegrass plus whole cottonseed (WCS) supplemented at 0.65% BW daily. Mean forage mass (kg DM/ha) was greatest in Yr 2, intermediate in Yr 1 and least in Yr 3, and greater for NFERT, CC, DDGS and WCS than AC. Mean ADG (kg/d) was greater in Yr 2 and 3 than Yr 1, and greater for NFERT, DDGS and WCS than CC and AC. Total gain (kg/ha) was greater in Yr 1 and 2 than Yr 3, and was greater for NFERT, DDGS and WCS than CC and AC. Mean stocking density (steers/ha) was greatest in Yr 2, intermediate in Yr 1 and least in Yr 3, and was greater for NFERT, DDGS and WCS than CC and AC. Steer-grazing-d/ha were greatest in Yr 1, intermediate in Yr 2, and least in Yr 3, and greater for NFERT, DDGS and WCS than CC and AC. Results are interpreted to mean that annual ryegrass pasture supplemented with either DDGS or WCS supported ADG, stocking densities, grazing-days/ha and total gain/ha that were similar to or greater than annual ryegrass amended with N fertilizer or interseeded with annual legumes.

[1]Phillip A. Gunter, Mary K. Mullenix, and Russell B. Muntifering Upchurch Hall Auburn University, AL 36849, Lance C. Burdette 4725 Co Road Shorter, AL 36075

EVALUATION OF OCTOBER VS NOVEMEBER PLANTING DATE ON SPRING DM YIELD OF FIVE WINTER ANNUAL FORAGES

T.A. Davis, K.J. Simon, J.A. Jennings, P.A. Beck, D. Philipp[1]

Winter annual forages are often planted for winter grazing, but producers typically wait for cool weather and fall rains before planting, typically in October. When sufficient rainfall does not, producers may not plant winter annuals or plant after dates that are recommended. This study compared October vs. November planting dates for spring forage production of five winter annual forages. The following forages were planted at the SWREC at Hope, AR: winter oat (Coker 227), rye (Elbon), wheat (AGS 2027), and ryegrass (Marshall and Nelson) at 120 lb/acre for oat, wheat, and rye, and 25 lb/acre for ryegrass. Plots were no-till drilled on October 6 and November 17, 2017 in bermudagrass sod suppressed with 1 pt/acre glyphosate. Fertilization was according to soil test recommendations for winter annual forages. Plots were harvested on March 27 and April 27, 2018. October-planted forages tended to have higher total DM than November-planted forages, but differences were not large. October treatments were at a more advanced stage of maturity in March than November treatments. Forage production was greater during April for the second harvest for the November treatments. This result indicates that to help ensure forage availability from winter annuals in March, fields should be planted in October. Winter annuals planted in November had delayed DM production in spring. To help extend the forage production period during spring, a portion of forage acres could be planted in both October and November to take advantage of the early and delayed maturity of each planting date.

[1]County Extension Agent-Agriculture, Univ. of Agriculture, Pike County Cooperative Extension Service, Murfreesboro, AR (tadavis@uaex.edu); Program Associate Animal Science/Forages, Little Rock, AR, (ksimon@uaex.edu); Professor and Extension Forage Specialist, Little Rock, AR, (jjennings@uaex.edu); Professor - Animal Science, Hope, AR, (pbeck@uaex.edu) and Associate Professor - Animal Science, Fayetteville, AR (dphilipp@uaex.edu)

FERTILITY AND CUTTING HEIGHT AFFECT ORCHARDGRASS PERSISTENCE

J.A. Tredaway, C.C. Mitchell, Jr., and D.M. Ball[1]

Orchardgrass is an important forage grass in much of the United States. Among other attributes, it provides good forage yield, has a high nutritive content, and is compatible with forage legumes. It is a perennial, but many forage-livestock producers believe that persistence of this grass has declined within the past two decades. A combination of any of several factors may be involved, but fertility and cutting height are generally believed to be particularly important. Northeast Alabama is on the edge of the area in which orchardgrass is well adapted, and thus is a location in which the grass is especially sensitive to stress factors. A study conducted on the Sand Mountain Research and Education Center was designed to investigate the impact of these factors on orchardgrass persistence. It included 16 fertility treatments (most involving various rates of nitrogen, phosphorus, and potassium), plus mowing heights of 2- and 4-inches that were imposed on each fertility treatment. Both soil fertility and cutting height had direct and important impacts on orchardgrass persistence. With regard to soil fertility, low levels of soil potassium were particularly detrimental. A 4-inch cutting height, was much more conducive to persistence than a 2-inch cutting height. In this test there were no interactions between fertility treatments and cutting heights.

[1]Former Superintendent, Sand Mountain Research and Education Center, Crossville, AL; and Professors Emeritus, Auburn University, respectively.

FILLING THE FALL FORAGE GAP WITH SUMMER ANNUAL FORAGES

J. A. Jennings, K. J. Simon, D. Philipp, J. V. Skinner, and G. Lorenz[1]

A major gap for growing high quality forage occurs from August through October in the southeast U.S. Early planting of winter annual forages incurs risk of dry, hot weather typical of this period. Annual ryegrass and small grains may establish, but perform poorly until the cooler weather of late October. Summer annual forages are adapted to higher temperatures and may offer some drought tolerance. This study determined forage production potential of selected summer annual forages during the late summer. In late August, 2017, pearl millet, sorghum-sudan, corn, soybean, cowpea, and lablab were planted. In 2018, pearl millet, browntop millet, sorghum-sudan, and corn were planted. Plots were harvested in mid-October both years. Good soil moisture was present at planting in 2017, but no rain fell for 26 days after planting, and September average high temperatures were higher than August. Pearl millet and sorghum-sudan produced over 3,600 lb DM /acre DM. Two corn varieties produced 1,300 to 1,500 lb DM/acre due to low plant populations reduced by wireworm. Cowpea, lablab, and soybean yield was below 600 lb DM/acre. In 2018, excellent rainfall resulted in excellent stands of all forages. Pearl millet and sorghum-sudan yield exceeded 3,000 lb DM/acre, and corn yield ranged from 2,000 to over 4,000 lb DM/acre. Yield of pearl millet and sorghum-sudan was similar for both years even under drastically different rainfall conditions. These studies indicate that summer annual forages have great potential to fill the forage gap in late summer and early fall before winter annual forages produce DM.

[1]Professor, Forage Specialist, Little Rock, AR (jjennings@uaex.edu); Program Associate – Animal Science/Forages, Little Rock, AR, (ksimon@uaex.edu); Associate Professor – Animal Science, Fayetteville, AR (dphilipp@uaex.edu); Resident Director, Arkansas Agricultural Research and Extension Center, Fayetteville, AR (jskinner@uark.edu); Distinguished Professor and Associate Entomology Department Head, Lonoke AR (glorenz@uaex.edu).

FINE-TUNING TEMPERATURE REQUIREMENTS FOR CLOVER SEED GERMINATION

Lisa Baxter, Tim Grey, Jennifer Tucker, and Dennis Hancock[1]

Clovers (*Trifolium* spp.) are an important component for improving the nutritive value of forages in forage-livestock grazing systems in the Southeastern US. Temperature significantly affects the time and rate of clover germination. The objectives of this trial were to determine the (1) temperature and time required to reach maximum germination and (2) the range in which germination will occur for six coated clover varieties adapted to the Southeast. Seed germination and vigor were evaluated over two thermal gradients ranging from 30 to 118.0 ºF where germination was counted every 24 hours for seven consecutive days. Three-parameter logistic growth models were used to find the minimum and maximum temperatures at which these varieties could maintain 80% germination. Lorentzian distribution models were used to establish the temperature and time (hours) to maximum germination. These six varieties maintained at least 80% germination from 41 to 83 °F on average, although some level of germination was observed beyond these extremes. The range at which theses clovers could germinate was much wider than reported in previous literature. Excluding Barduro, maximum germination occurred at 34.2 to 52.8 h after planting at optimum temperatures (52 to 63°F). Barduro clover germinated so rapidly in the warmer temperature range that the full data set could not be used as it did not conform to the same regression. These data will be useful in updating planting recommendations for producers in the Southeast.

[1]Post-doctorate, Crops and Soils Science Dept., University of Georgia, Tifton, GA (baxterl@uga.edu); Professor, Crops and Soils Science Dept., University of Georgia, Tifton, GA, (tgrey@uga.edu); Assistant Professor, Animal and Dairy Science Dept., University of Georgia, Tifton, GA, (jjtucker@uga.edu); Forage Extension Specialist, Crop and Soils Science Dept., University of Georgia, Athens, GA, (dhancock@uga.edu).

FORAGE ALTERNATIVES AND MANAGEMENT WILL EXTEND THE GRAZING SEASON IN THE GULF COAST REGION

G. Scaglia[1]

For most livestock producers, extending the grazing season or otherwise filling gaps in forage availability to reduce stored feed needs should be a high-priority objective. Grazing standing forages benefits the environment, weather is less of a concern, better forage leads to better animal performance, requires less labor, and lowers expenses. In livestock budgets, stored feed typically accounts for 25% or more of the cost of production. In general, the less hay needed, the more cost-efficient the operation. In the Gulf Coast region, it is possible to stockpile summer perennials with relative success. Brassicas and cereal grains can provide early grazing during the fall transition period. On the other hand, white and red clovers have proven to provide grazing in late spring until mid-summer. In three consecutive years, 36 newly weaned steers (BW=524±21 lb) were allotted to 3 treatments: 1) bermudagrass hay; 2) stockpiled bermudagrass, cv. 'Jiggs'; and 3) 'cocktail' pasture mix of forage radish, kale, turnips, oats, annual ryegrass, white and red clover. Calves in treatments 1 and 2 were supplemented daily with 2 lb/head of soybean hulls pellets due to low forage nutritive value. On average, steers grazing the 'cocktail' pasture mix gained more (2.21 lb/d) than those consuming hay (1.32 lb/d) while those on stockpiled bermudagrass gained the least (0.98 lb/d). Nutritive value of stockpiled bermudagrass significantly declined due to rainfall and first frost therefore affecting steers' gains. Mixed forages ('cocktail') provide greater nutritive value and extended the grazing season (fall and spring) for weaned calves with lowest costs of production.

[1]Professor, LSU AgCenter Iberia Research Station, Jeanerette, LA (gscaglia@agcenter.lsu.edu).

FORAGE PRODUCTION AND SWARD CHARACTERISTICS OF NATIVE WARM-SEASON GRASS MIXTURES UNDER DIFFERING NITROGEN FERTILITY RATES

C.G. Nichols[1], M.L. Marks, K.M. Mason, M.K. Mullenix, S.L. Dillard, J. Yeager, and R.B. Muntifering

Native warm-season grasses (NWSG) may provide an alternative forage system to endophyte-infected tall fescue in the Black Belt prairie soil region of Alabama and Mississippi during the summer months, but evaluation of herbage production potential and persistence under grazing is needed. Our objective was to evaluate nitrogen (N) fertilizer rate effects (0 or 60 lb N/acre; n = 3, five acre paddocks/treatment) on herbage mass and sward characteristics of a mixture of big bluestem, little bluestem, and indiangrass in the Black Belt region of Alabama. Paddocks were continuously stocked with four weaned Angus x Simmental beef heifers from early June to late August 2018 (71 days of grazing). Additional put-and-take heifers were used to manage forage to a target of 15 in. Forage mass and sward heights were collected every two weeks. Ground cover, light interception, and botanical separations were measured at the beginning and end of the study. Nitrogen-fertilized NWSG had greater forage production than the unfertilized control (1,926 vs. 1,600 lb DM/acre, respectively). Ground cover and botanical estimates illustrated that weed competition was greatest at the beginning of the grazing season following fertility application, which may have implications in terms of species competition and sward persistence over time. Canopy light interception decreased by 47% for N-fertilized vs. unfertilized NWSG (-17%) during the grazing season. Year 1 data illustrates that NWSG systems may provide a viable grazing system in the summer months under reduced N inputs, although more long-term persistence measures will be collected in subsequent years of the trial.

[1]Graduate Research Assistant, Auburn University Department of Animal Sciences, Auburn, AL (cgn007@auburn.edu).

FORAGE QUALITY AND CATTLE PERFORMACE DURING THE BACKGROUNDING STAGE ON TALL FESCUE PASTURES INTERCROPPED WITH SUNN HEMP

Joshua Tooley[1] and H.D. Naumann[2]

Cool-season grasses comprise a majority of pasture swards in the northern transition zone. Productivity of cattle grazing during the backgrounding stage becomes limited due to decreased forage nutritive value and yield during the summer. Intercropping an annual warm-season legume like sunn hemp (SH) may improve forage nutritive value resulting in greater cattle performance. The objectives of our experiment were to compare forage nutritive value of cool-season grass pastures intercropped with SH to a tall fescue pasture, and differences in animal performance and production in those systems. Treatments imposed on established novel-endophyte tall fescue pastures included: TF+N, with nitrogen applied at 40 lb/acre in June and 80 lb/acre in August, and TF+SH, with SH planted post-hay or post-grazing from June to August (40 lb/acre). All treatments were compared to unfertilized tall fescue (TF-N). Weaned cattle weighing 675 pounds, grazed rotationally for 120 days starting in July. TF+SH and TF+N resulted in greater crude protein than TF-N during October and November of 2017 and the spring of 2018. TF+SH neutral detergent fiber concentrations were two percent less than all other treatments. Average daily gain (AGD) was similar for cattle grazing all treatments. However, cattle on TF+N had the greatest productivity (210 lb gained/acre) followed by TF+SH (168 lb gained/acre). Our results suggest intercropping SH and nitrogen fertilization positively influences forage nutritive value compared to TF-N. Greater productivity observed in TF+N and TF+SH may have been due to greater forage quality, regrowth, and forage accumulation.

[1]University of Missouri, Division of Plant Science (JATC65@mail.missouri.edu)
[2]University of Missouri, Division of Plant Science (naumannHD@ missouri.edu)

HEIFER DEVELOPMENT & FORAGE EVALUATION IN SILVOPASTURE SYSTEMS

Sarah Jane Thomsen, Gabriel Pent, & John Fike[1]

Silvopasture is the intentional integration of trees with forages and livestock. A silvopasture was developed from a thinned timber stand in Blackstone, Virginia, and seeded with a cool-season forage mixture. Treatment pastures were an open pasture, a thinned pine silvopasture, a thinned hardwood silvopasture, and a cleared and replanted new silvopasture. Cattle were introduced in 2017 and rotationally stocked within each pasture according to forage availability. Objectives were to determine the forage availability, forage nutritive value, and the performance of heifers in silvopasture and open pasture systems. Forage availability was similar between treatments in 2017 (5040 lb/ac); but lowest in the hardwood silvopastures (2980 lb/ac) compared to the other treatments (3240 lb/ac) in 2018. There was no significant difference in crude protein between pastures (12.7%). In 2017, ADG of the heifers were similar in all the treatments (1.3 lb/day). In 2018, ADG was lowest in hardwood silvopastures (0.81 lb/day) compared to other treatments (1.0 lb/day). Remote temperature loggers were used to collect vaginal temperatures of the heifers over eight days in 2018. Silvopasture heifers had an average core temperature of 102.9 °F from 2-5 PM, while heifers in the open pastures had an average temperature of 104.0 °F. A drone with a thermal camera was used to collect external hide temperatures in the morning and afternoon. Heifers in the silvopastures had lower heat loads in the afternoon. Silvopastures provide an opportunity to improve the welfare of grazing livestock in the summer, while improving the productivity of cool-season grass pastures.

[1]Graduate Research Assistant, Virginia Tech Southern Piedmont AREC, Blackstone, VA (tsarah95@vt.edu); Ruminant Livestock System Specialist, Virginia Tech SPAREC, Blackstone, VA (gpent@vt.edu); Forage Specialist, Virginia Tech, Blacksburg, VA (jfike@vt.edu).

IMPACT OF PERENNIAL WARM-SEASON GRASS FORAGES ON PROFITABILITY OF BEEF CATTLE OPERATIONS IN THE FESCUE BELT

K.A. Brazil, P.D. Keyser, A.P. Griffith, C.N. Boyer, J.T. Mulliniks, G.E. Bates[1]

Profitability of beef cattle operations across the southeastern United States is heavily influenced by forage cost, quality, and quantity. Within the tall fescue belt, forage production during summer months is limited and animal productivity may be further compromised by fescue toxicosis. We developed an enterprise-level economic model to determine if incorporating perennial warm-season grasses into existing cool-season grass forage systems for beef cow-calf production is a profitable management alternative. A sensitivity analysis determined the effects of summer drought of varying lengths (2 to 10 weeks) and purchased feed costs (early-bloom orchard grass hay; costs based on USDA AMS Virginia Hay Report) on profitability. Simulations include spring and fall-calving herds grazing tall fescue only, tall fescue plus switchgrass, and tall fescue plus bermudagrass. Pasture costs (variable and fixed) included forage crop establishment (pro-rated over ten years) and annual management, winter and summer supplemental feed (hay only), and interest (6%) on operating capital. Revenues consisted of sale of calves and cull animals. Cattle prices were based on USDA AMS data. Calving rates and weaning weights were assigned based on representative data in the peer-reviewed literature. The model determined which forage system and calving season combination will benefit the most, if at all, from incorporating perennial, warm-season grasses into existing cool-season forage systems under simulation.

[1]Graduate Research Asst, University of Tennessee (UT), Knoxville, TN (kbrazil@utk.edu); Professor, UT, Knoxville, TN (pkeyser@utk.edu); Asst Professor, UT, Knoxville, TN (agriff14@utk.edu); Assoc Professor, UT, Knoxville, TN (agriff14@utk.edu); Asst Professor, University of Nebraska, Lincoln, NE (travis.mulliniks@unl.edu); Professor, UT, Knoxville, TN (gbates@utk.edu).

IMPACT OF POTASSIUM APPLICATION AND HARVEST REGIME IN ALFALFA YIELD, FORAGE QUALITY AND STAND PERSISTENCE IN SOUTH GEORGIA.

B.M. Thinguldstad*, J.J. Tucker, J.R. Segers, and L.L. Baxter

Potassium (K) fertilization and harvest timing can both impact stand life of alfalfa (*Medicago sativa* L.). Management recommendations are generated from northern alfalfa varieties. However, high temperatures and humidity in the Southeast necessitate the use of different varieties. The objective of this trial is to determine the impact of potassium fertilization and harvest regime on forage yield and stand persistence of alfalfa in the Southeastern US. This study was conducted on a two-year-old stand of 'Bulldog 805' alfalfa planted fall 2015 at the Coastal Plain Experiment Station in Tifton, GA. Plots were randomly assigned using a split block design. Plots were harvest treatment: bud, and 10, 30 and 50% bloom stage and within each harvest treatment were subplots of K fertilization rate: 0, 60, 90, 120 and 150 lbs/acre split-applied 3 times throughout the season. At harvest, plots were assessed for ground cover. Fifty shoots were collected to evaluate leaf:stem and mass:shoot ratio, and yield was determined using a Swift forage harvester. Grab samples were collected for nutritive value analysis using near-infrared reflectance spectroscopy (NIR). Harvest treatment had an effect on yield, with 10% bloom being greater than bud stage (1179 lbs/ac and 985 lbs/ac, respectively; $P = 0.04$). Crude protein and TDN levels were the greatest in bud stage and lowest in 50% bloom (22.8% and 21.2%, $P < 0.01$ for CP; 65% and 53% for TDN, $P < 0.01$; respectively). Funding for this project was provided by USDA-NIFA-AFRP grant #2016-70005-25653.

[1]Graduate student, The University of Georgia, Tifton Campus (brittat@uga.edu); Assistant professor, The University of Georgia, Tifton Campus (jjtucker@uga.edu); Assistant professor, The University of Georgia, Tifton Campus (jacobs@uga.edu); Post-Doctoral Research Associate, The University of Georgia, Tifton Campus (baxterl@uga.edu)

IMPACT OF POULTRY LITTER APPLICATION ON YIELD AND QUALITY OF ALFALFA GROWN IN MISSISSIPPI

J.B. Rushing[1], R.W. Lemus[2], J.G. Maples[3], and J.C. Lyles[4]

Alfalfa production in the Deep South has increased over the past 10 years due to varietal development and improved management systems. This increase in production had generated the need for more economical and sustainable nutrient management plans in alfalfa production, particularly in areas where poultry litter (PL) is readily available. A field trial was conducted in Starkville and Newton, MS during the 2018 growing season to compare poultry litter and synthetic fertilizer applications. The trial consisted of a randomized split-plot arrangement with 4 fertilizer treatments [(1) 1 and (2) 2 tons PL/acre; (3) N-P-K based on PL analysis; (4) P-K based on PL analysis] and three varieties ([(1) 'Bulldog 505'; (2) 'Bulldog 805'; (3) 'AlfaGraze 600RR'] and was replicated 4 times at each location. Dry matter (DM) yield, stand persistence, and forage quality was assessed at each harvest. The Starkville site was abandoned due to poor drainage and weed pressure. Five harvests were conducted at Newton. Bulldog 505 and AlfaGraze 600RR had the greatest cumulative yield with 10,512 and 9,464 lb/acre, respectively. No differences were observed between fertility treatments at the Newton site. Crude protein (CP) ranged from 20.8 to 26.5% across all five harvests. Relative forage value (RFV) was affected by variety and fertilizer treatment. Bulldog 505 had the lowest seed cost per pound of mean DM ($0.0099). Based on a single site year of data collection, Bulldog 505 with a 1 ton PL/acre fertilizer application generated the greatest amount of DM with the least cost.

[1]Assistant Research/Extension Professor; Coastal Plain Branch Experiment Station, Newton, MS
[2]Forage Extension Specialist; Department of Plant and Soil Sciences, Mississippi State, MS
[3]Assistant Extension Professor; Department of Agricultural Economics, Mississippi State, MS
[4]Research Associate I; Coastal Plain Branch Experiment Station, Newton, MS

IMPACTS OF REDUCED LIGNIN ALFALFA ON FORAGE DIGESTIBILITY IN HORSES

A.M. Grev[1], A.S. Reiter[1], M.R. Hathaway[2], C.C. Sheaffer[3], and K.L. Martinson[2]

Reduced lignin alfalfa has the potential to provide a higher quality forage source for livestock by improving forage digestibility. This study was conducted to evaluate apparent dry matter digestibility (DMD), crude protein digestibility (CPD), neutral detergent fiber digestibility (NDFD), and fecal particle size (FPS) when feeding reduced lignin and reference alfalfa hays to adult horses. In 2017, reduced lignin ('54HVX41') and reference ('WL355.RR') alfalfa hays were harvested in Minnesota at the late-bud stage. Alfalfa varieties were similar in crude protein (20%) and neutral detergent fiber (43%) content. Acid detergent lignin concentrations were lower for '54HVX41' (7.3%) compared to 'WL355.RR' (8.1%; $p = 0.002$). Dietary treatments were fed to six adult horses in a crossover study. Experimental periods consisted of a 9-d dietary adaptation phase followed by a 5-d total fecal collection phase, during which horses were housed in individual boxstalls and manure was removed on a continuous 24-h basis. At 12-h intervals, feces were thoroughly mixed, subsampled in duplicate, and used for apparent digestibility and FPS analysis. Dietary treatments were similar in dry matter intake (1.6% bodyweight) and time to consumption (604 min). Apparent DMD was greater for '54HVX41' (64%) compared to 'WL355.RR' (62%; $p = 0.048$). Apparent CPD, NDFD, and FPS did not differ between dietary treatments, averaging 78%, 45%, and 0.89 mm, respectively. These results indicate an improvement in DMD for reduced lignin alfalfa hay when fed to adult horses, with no change in forage consumption or FPS.

[1]Graduate Research Assistant, Department of Animal Science, University of Minnesota (grevx004@umn.edu; reit0162@umn.edu); [2]Professor, Department of Animal Science, University of Minnesota (hathaway@umn.edu; krishona@umn.edu); [3]Professor, Department of Agronomy and Plant Genetics, University of Minnesota (sheaf001@umn.edu)

MANAGING THE SUGARCANE APHID IN FORAGE SORGHUM: CULTIVAR X APHICIDE INTERACTIONS

C.D. Teutsch, R.T. Villanueva, Z.J. Vilora, G.L Olson, and S.R. Smith[1]

Forage sorghum (*Sorghum bicolor* (L.) Moench) could provide a drought tolerant alternative to corn (*Zea mays* L.) for silage production in the upper Southern U.S. However, a new pest of sorghum, the sugarcane aphid (*Melanaphis sacchari* (Zehntner) (Hemiptera: Aphididae), could restrict its use. The objectives of this study were to document of tolerance of forage sorghum cultivars to the sugarcane aphid and to evaluate the efficacy of an aphicide on these same cultivars. In May 2018, cultivars were planted at University of Kentucky's Research and Education Center at Princeton. The experimental design was a randomized complete block with split plot treatment arrangement and four replications. Sivanto (Flupyradifurone) aphicide was applied at a rate of 6 oz/A on 24-Aug-18 as a drench. In late September, plots were rated for sugarcane aphid damage using a scale of 1 to 9, with 1 being little or no damage and 9 being severe damage. Sugarcane aphid damage was less in plots treated with the aphicide ($P < 0.01$), although the range of the difference was smaller than anticipated (6.3 versus 9.7). Cultivars also differed in sugarcane aphid damage ($P < 0.10$), with a range of 5.7 to 8.3. There was no aphicide x cultivar interaction for the damage rating ($P > 0.70$). The use of aphicides will likely be required in the short-term for management of sugarcane aphids in forage sorghum. However, selection of varieties with increased tolerance to the sugarcane aphid may provide a simple and cost effective approach to management in the future.

[1]Associate Extension Professor, Dept. of Plant and Soil Sciences, Univ. of Kentucky, Lexington (chris.teutsch@uky.edu); Assistant Extension Professor, Dept. of Entomology, Univ. of Kentucky, Lexington; Research Analysist, Dept. of Entomology, Univ. of Kentucky, Lexington; Agriculture Research Specialist Sr, Dept. of Plant and Soil Sciences, Univ. of Kentucky, Lexington; Extension Professor, Dept. of Plant and Soil Sciences, Univ. of Kentucky, Lexington.

NITROGEN APPLICATION ON DIVERSE SUMMER ANNUAL FORAGE MIXTURES

K.M. Mercier, C.D. Teutsch, S.R. Smith, E.L. Ritchey, K.H. Burdine, and E.S. Vanzant[1]

Increasing biodiversity has often been linked to increased productivity, especially when including legumes. However, in annual systems, legumes may not supply nitrogen to associated plants during the growing season. An experiment was conducted at Princeton, KY (Zanesville silt loam with a fragipan), and Lexington, KY (Bluegrass-Maury silt loam) to evaluate the effects of nitrogen (0, 50, 100, 150, and 200 lb N/A) application on summer annual forage mixtures of varying botanical diversity [sorghum-sudangrass monoculture, simple mixture (three species), and complex mixture (12 species)]. Plots were harvest three times in 2018. In Lexington, both mixtures out-yielded the monoculture (5210 vs. 4740 lb DM/A, $P < 0.001$), while in Princeton, forage mixture had no effect on yield (3560 lb DM/A, $P > 0.5$). Applied nitrogen increased yields of Lexington plots (5170 vs 4250 lb DM/A, $P < 0.001$), while Princeton plots showed a positive linear trend in response to N (2670-5000 lb DM/A, $P < 0.001$). At both locations, mixture had no effect on crude protein (CP) or total digestible nutrients (TDN) ($P > 0.09$). Increasing N application increased CP in both Lexington (15-18%, $P < 0.001$) and Princeton (8-11%, $P < 0.001$) and TDN at both locations (Lexington: 60-62%, $P < 0.001$; Princeton: 58-60%, $P < 0.001$). Results from this study indicate that summer annual forage mixtures have the potential to out-yield monocultures, and up to 200 lb N/A can help improve forage nutritive characteristics; however, results may be dependent upon soil type and weather.

[1]Graduate Student, University of Kentucky Research and Education Center at Princeton (UKREC), Princeton, KY (kelly.mercier@uky.edu); Extension Associate Professor, UKREC, Princeton, KY (chris.teutsch@uky.edu); Extension Professor, University of Kentucky (UK), Lexington, KY (raysmith1@uky.edu); Extension Associate Professor, UKREC, Princeton, KY (elritc2@uky.edu); Associate Extension Professor, UK, Lexington, KY (kburdine@uky.edu); Associate Professor, UK, Lexington, KY (evanzant@uky.edu).

NUTRITIVE VALUE AND DRY MATTER YIELD OF REDUCED-LIGNIN ALFALFA IN GRASS MIXTURES

S.R. Smith, J.H. Cherney, C.C. Sheaffer, D.J.R. Cherney, and M.S. Wells[1]

New reduced-lignin alfalfa varieties and higher nutritive value grass species are now being marketed and provide an opportunity for farmers to increase forage quality and profitability. This research assessed the yield and nutritive value of reduced-lignin alfalfa (Hx14376) vs. standard alfalfa (WL 355.RR) in monoculture and binary mixtures with three perennial grasses (Fojtan festulolium, BARFpF32 meadow fescue, and Dividend VL orchardgrass). The forages were seeded into prepared seedbeds on University farms in the spring of 2016 [mid-April Ithaca, NY (Cornell), late-March Lexington, KY (Univ. of KY) and late May Rosemount, MN (Univ. of MN)]. Alfalfa varieties were Roundup Ready[R] with 4 FD rating and harvests taken at the bud and flower stages (2016-spring 2018). Reduced-lignin alfalfa averaged 14.1% less acid detergent lignin and 5.9% greater neutral detergent fiber digestibility, compared with the standard variety. Seasonal forage yield was slightly higher for the standard alfalfa vs. the reduced-lignin variety. Mixed alfalfa-grass stands had similar yields to pure alfalfa. Flower stage harvest resulted in increased yields vs. bud stage harvest in a four cut system, but had reduced yields when four cuts were taken at the flower stage vs. five cuts in the bud stage. While the response of alfalfa across regions was relatively consistent, grass species were inconsistent across regions for grass percentage, yield and quality. In conclusion, reduced-lignin alfalfa can provide higher quality alone or in grass mixtures, but evaluations need to be conducted on a regional basis.

[1]S.R. Smith, University of Kentucky (raysmith1@uky.edu); J.H. Cherney and D.J.R. Cherney, Cornell University; and C.C. Sheaffer and M.S. Wells, University of Minnesota.

PASTURELAND NRI FINDINGS IN ILLINOIS

Roger L. Staff[1]

For several years Illinois has participated in the Pastureland Natural Resource Inventory (NRI), which is a survey completed annually on grassland sites across the nation using 15 protocols. The grassland sites include pasture, permanent hay lands, idle grassed areas, wetland WRP sites and some continuous CRP practices, as Pollinator Habitat, and Filter Strips. The largest protocol to complete is the Line Pont Intercept. In 2018, 16 counties were identified with segments consisting of 50 points for data collection; 11, or 22%, had data collected from those points. Some sites are in cropland, timbered areas, located on farmsteads or urban areas, and some had access denied. Of 11 points with data collected, only two were pastures, and of those two, only one had any legumes present (6.34% of the vegetation identified). In addition to the legumes, I looked at the number of Pervasive Plants present from the NRCS Illinois State list. There were 17 Pervasive Plants identified, or 4.99% of the stand. Also, we identified 30 Pollinator Plants that made up 3.64% of the plants sampled. In 2017, 16 counites had 48 points reviewed and data was collected on 11 (22.9%). The four counties with pastures had 1.29%, 21%, 28.6%, and 28.9% legumes. Pervasive Plants represented 5.4% of all plants identified and Pollinator Plants represented 4.69%. In 2016, 14 counites had 42 points reviewed and data collected on 18 points, for a 42.8% ratio. Of those 18 points, there were 4 that had pastures with a legume percent of 0.34%, 8.28%, 6.45%, and 2.76%. These were very low! The Pervasive Plants identified were 3.94% of the total. The Pollinators Plants were 2.4% of the identified list.

[1]USDA-NRCS Natural Resource Inventory Statistics, (Roger.Staff@il.usda.gov). 2015 NRI Summary Report & NRI Pastureland Resource Assessment, June 2018.

PERFORMANCE OF STOCKERS GRAZING DIVERSE SUMMER ANNUAL FORAGE MIXTURES

K.M. Mercier, C.D. Teutsch, S.R. Smith, E.L. Ritchey, K.H. Burdine, and E.S. Vanzant[1]

Retaining calves on summer pasture provides an opportunity to add extra gain before selling at a more opportune late-summer market. The dominant perennial cool-season forages in the Mid-South often have insufficient quality and yield to support desired summer gains. In contrast, summer annual forages have both improved production and nutritive value during the summer months. In 2017 and 2018, a study was conducted near Princeton, KY, where calves (725 lb in 2017 and 806 lb in 2018) grazed one of three summer annual forage treatments: 1) sorghum-sudangrass monoculture, 2) simple mixture (three species), and 3) complex mixture (12 species). Due to differences in grazing management, results varied by year ($P < 0.04$). In 2017, calves grazing the simple mixture and the monoculture gained 1.74 lb/day while calves grazing the complex mixture gained 1.46 lb/day ($P < 0.03$). In 2018, no differences in average daily gain were detected among treatments and the calves only gained 0.02 lb/day. The low average daily gains in 2018 were likely due to higher nutritional demand of heavier calves and the lower nutritive value of more mature forage. Keeping summer annual forages in a vegetative state is paramount to maintaining adequate gains during the summer months. The yield and nutritive value of the mixtures will be reported.

[1]Graduate Student, Univ. of Kentucky Res. and Educ. Center at Princeton (UKREC, Princeton (kelly.mercier@uky.edu); Exten. Assoc. Prof., UKREC, Princeton, KY; Exten. Prof., Univ. of Kentucky, Lexington; Exten. Assoc. Prof, UKREC, Princeton, KY; Assoc. Exten. Prof., University of Kentucky, Lexington; Assoc. Prof, University of Kentucky, Lexington, KY.

PROFITABILITY CENTERS ON MANAGED GRAZING DAIRIES

T. Cadwallader, J. Jewett, L. Paine, K. Solberg, J. Paulson, and D. Gucker[1]

The objective of the project was to determine input categories that grazing dairies use to reduce overall costs and improve net profit. This was part of a grant from the United States Department of Agriculture Risk Management Education Partnership Program. In order to determine the cost of production portion of the project, seven dairy farms from the upper Mississippi watershed (Minnesota, Wisconsin, Iowa and Illinois) that had a history of effectively using Management Intensive Grazing (MiG) agreed to share their 2017 cost of production information. The average of those seven MiG dairy farms was then compared to dairy farm benchmark numbers of dairy farms of similar size from one of the benchmarking sources commonly used in the upper Midwest, the Wisconsin Agricultural Financial Advisor (AgFA) program managed by the University of Wisconsin Center for Dairy Profitability (CDP), https://cdp.wisc.edu/agfa-farmbench. The results showed our sample MiG dairies were 10 times more profitable than AgFA benchmark dairies. It was found that this was due to savings in the following areas: crop production costs, animal health and herd longevity, and equipment costs.

[1]Extension Educator, University of Illinois, Decatur, IL 62521 (dgucker@illinois.edu).

REEVALUATING NITRATE TOXICITY POTENTIAL WHEN GRAZING ANNUAL FORAGES

M.E. Lenz, R. J. Kern, J. K. Farney and M.E. Drewnoski[1]

Of the annual forages (n=1957) submitted to a commercial laboratory (Ward laboratory, Kearney, NE) in 2016-2017, 48% of fresh brassica samples, 23% of fresh annual grasses, and 5% of dry annual grasses exceeded 2,100 ppm NO_3-N and would have been considered at risk for causing toxicity according to most recommendations. An online survey distributed through the Nebraska Extension "BeefWatch" newsletter and advertised at extension meetings in NE and KS found that 70% of respondents (n = 80/115) were concerned about nitrates in annual forages, while only 38% experienced an issue. Only 38% of producers "frequently" tested annual forage pasture and 53% "frequently" tested hay. Producers tended to experience nitrate toxicity more often when grazing (31%) compared to feeding hay (21%). However, the increased incidence of toxicity in pasture is smaller than expected based on the large increase in the proportion of fresh forages that would be considered toxic in the commercial laboratory data set. An analysis of the nitrate content of seven late-summer planted annual forages grazed in fall by weaned calves, which had no observable signs of toxicity, found that 5 of the fields had nitrate concentrations 2 to 4 fold greater than the 2,100 ppm NO_3-N threshold. There are factors that may explain differences in toxicity risk for animals grazing annual forages compared to animals fed hay, and these should be considered when formulating recommendations. These data suggest a need to reevaluate the guidelines for risk of nitrate toxicity when grazing annual forages.

[1]Graduate Student, University of Nebraska, Lincoln, NE; Animal Scientist, Ward Laboratories Inc., Kearney, NE; Beef Systems Specialist, Kansas State University Southeast Research and Extension Center, Parsons, KS; Beef Systems Specialist, University of Nebraska, Lincoln, NE (mdrewnoski2@unl.edu)

RENOVATION STRATEGIES FOR TOXIC ENDOPHYTE-INFECTED TALL FESCUE FOR PROFITABILITY, ANIMAL AND AGRONOMIC PERFORMANCE, AND SOIL HEALTH

S. Ingram[1], M. Poore[1], A. Franzluebbers[2]

[1] Department of Animal Science, NC State University, Raleigh, NC
[2] USDA-ARS, Department of Crop and Soil Science, NC State University, Raleigh, NC

The negative impact of toxic endophyte-infected tall fescue (TF) on the livestock industry has long been established and is well documented, yet conversion of the 38 million TF acres to non-toxic endophyte infected (NE) has been minimal. More information relating to the entire production system during the renovation of TF pastures is needed. Therefore, three treatments were evaluated for profitability, forage production and value, animal performance and soil health. Treatments included: control (C), renovation to novel endophyte-infected tall fescue (NE) after one season of a single specie cover crop beginning in spring 2018 (1-SM), renovation to NE after three seasons of a single specie cover crop beginning in spring 2017 (3-SM), and renovation to NE after three seasons of a multi-specie cover crop beginning in spring 2017 (3-CM). Soil samples were taken in March 2017 to establish baseline soil health measurements for each treatment. After one season of data collection in the summer of 2017, 3-SM forage yield was greater for the first two sampling dates of the grazing period (P = 0.0078, 4025 lb/a and P = 0.0019, 4114 lb/a; respectively) in comparison to 3-CM, C, 1-SM (2922, 2385, 2607 lb/a for the first sampling and 3358, 2918, 3031 lb/a for the second sampling, respectively). No difference (P = 0.5677) in forage yield was noted for the third sampling date but for the last sampling date, 3-CM forage yield was greater (P = 0.0013, 2615 lb/a) than 3-SM, C and 1-SM (1832 lb/a, 1189 lb/a, 1178 lb/a, respectively). Gain per acre was greater (P <0.001) for 3-SM and 3-CM compared to the control and 1-SM (163, 146, -21, -40 lb/a, respectively). Establishment costs for 3-CM ($128/acre) and 3-SM ($114/acre) were higher (P <0.001) than 1-SM and C ($63/acre). Net return was higher (P <0.001) for 3-SM ($41/acre) in comparison to 3-CM ($10/acre), which was higher (P <0.001) than 1-SM and C ($-101/acre, $-83/acre, respectively). Data suggests the two cover crops, 3-SM and 3-CM, improved system productivity in comparison to the two treatments that remained in TF.

RESULTS OF 2017 FERTILIZER OPTIONS FOR STOCKPILING COOL SEASON GRASS

M.A. Landefeld, D.F. Lima, J.S. McCutcheon, C.D. Penrose[1]

Abstract

Stockpiling cool season grass for grazing in the fall and winter is an option to extend the grazing season. The purpose of this study was to determine the effects of yield and quality by adding urea, urea with Agrotain®, and ammonium sulfate by applying them to cool season grass, primarily fescue and orchardgrass. There were three sites in Southeast Ohio, each was a randomized complete block design with four treatments (control, 100 lb/A urea, 100 lb/A urea plus Agrotain®,and 219 lb/A ammonium sulfate) and four replications of each treatment. The application date was July 31, 2017 and the plots were harvested on November 13, 2017. There were no statistical differences in CP, ADF, and TDN ($P < 0.05$). There were significant differences in yield between the control and all of the treatments, but not between the treatments. The three site average for the control was 2645 lb DM/acre; urea, 3322 lb DM/acre; urea + Agrotain®, 3494 lb DM/acre; and ammonium sulfate, 3278 lb DM/acre. Rainfall in the first 30 days from trial initiation in 2017 ranged from 3.2 to 3.3 inches, and the first significant rainfall (0.7-1.3 inches) was within 96 hours of initiation. Research will be continued to determine if any of the treatments provide significant differences in yield, CP, ADF, and TDN, and what effects rainfall after initiation of treatments has on yield and quality.

Introduction: Many livestock owners use the granular form of urea nitrogen during late summer and fall trying to increase forage growth for "stockpiled" forage. Livestock are then allowed to graze the "stockpile" at a later date when other forages no longer are growing or available. This practice extends the grazing season and reduces the need for higher priced stored feed. This was an identical study to one that was conducted in 2016 to replicate over time and was designed to determine the effects of using urea; adding a urease inhibitor (nitrogen stabilizer) product to urea, at the labeled rate, before applying the urea to the forage; and applying ammonium sulfate. The purpose of the study was to determine any difference in dry matter accumulation between treatments and detect changes in quality characteristics of the forages.

Materials and Methods: There were three locations (Monroe, Noble and Morgan Counties) with a randomized complete block design at each location with four (4) treatments, including a control, and four (4) replications of each treatment. Each plot was six feet by 20 feet. The fields

[1]Assistant Professor and Educator, Ohio State University Extension; Extension Educator, Ohio State University Extension; Associate Professor and Assistant Director, Ohio State University Extension; Professor and Educator, Ohio State University Extension, respectively.

were mechanically harvested prior to treatments to a height of three inches. The control plots received no urea (46-0-0), urease inhibitor, or ammonium sulfate (21-0-0). For the other treatments, a total of 46 lb/acre of nitrogen was used in each treatment in the following manner: 100 lb urea/acre; 100 lb urea/acre plus Agrotain® added at the labeled rate of one gal/ton of fertilizer; and 219 lb/acre ammonium sulfate which was applied on July 31, 2017. The plots were harvested on November 13, 2017 to a height of three inches above ground level utilizing 2' x 2' subsamples from each plot. Each subsample was weighed fresh, and then taken to a laboratory for forage analysis. Each of the 48 samples was quality tested for Crude Protein (CP), Acid Detergent Fiber (ADF) and Total Digestible Nutrients (TDN). Statistics were calculated using Proc Mixed in SAS 9.3. Model included treatment, farm, and treatment by farm, with the random variable of rep within farm.

Results and Discussion: Table 1 lists the three site average for yield and quality. More detail for each specific county result and background information is available in the Fertilizer Options for Stockpiling Cool Season Grass (2018) technical report available at corn.osu.edu. Dry matter (DM) yields averaged 2645 lb/acre for the plots where no nitrogen (N) was applied, 3322 lb/acre for the plots with 46 lb of N in the form of urea applied, 3494 lb/ acre for the plots with 46 lb of N in the form of urea applied, with Agrotain® added, and 3278 lb for the plots with 46 lb of N in the form of ammonium sulfate applied. There was a significant difference between the control and the treatments ($P < 0.05$) for yield, but not between the treatments. There was also a significant difference in CP between the control and the treatments, but not between the treatments. There were no significant differences with ADF and TDN.

Research by McInnes, Ferguson, Kissel, and Kanemasu (1986) has demonstrated that urea nitrogen can be susceptible to volatilization under dry conditions when no rainfall occurs to move the broadcast nitrogen (N) into the soil in a timely manner. However, rainfall after the treatments were initiated occurred within 96 hours of the start of the study (0.7 in. Monroe Co., 0.75 in. Noble Co., 1.3 in. Morgan Co.) reducing the potential to lose N to volatilization. For the month of August, rainfall at the three sites ranged from 3.18 in. to 3.2 in.

There were no statistical differences in CP, ADF, and TDN ($P < 0.05$). There were significant differences in yield between the control and all of the treatments for the three site average, but not between the treatments. The three site average for the control was 2645 lb DM/acre; urea, 3322 lb. DM/acre; urea+Agrotain®, 3494 lb DM/acre; and ammonium sulfate, 3278 lb DM/acre.

In the identical 2016 study, dry matter (DM) yields averaged 2627 lb/acre for the plots where no N was applied, 3144 lb/acre for the plots with 46 lb of N in the form of urea applied, 3459 lb/acre for the plots with 46 lb of N in the form of urea applied, with Agrotain® added, and 3609 lb/acre for the plots with 46 lb of N in the form of ammonium sulfate applied. There was a significant difference between the control and the treatments ($P < 0.05$) for yield, but not between the treatments. There was also a significant difference in CP between the control and the treatments, but not between the treatments. There were no significant differences with ADF and TDN.

Previous research conducted by Penrose (2015), and Landefeld (2016), showed a numerical increase in dry matter accumulation occurred using treatments over the control (urea alone and urea plus Agrotain®) even though there was no significant difference in the treatments at the *(P* < 0.05) for yield. There was a significant difference in crude protein between urea (8.53%) and urea plus Agrotain® (8.31%) when compared to the control (6.77%) in the 2014 study. However, in the 2015 study, there was a significant difference between urea plus Agrotain® compared to the control and urea only.

One needs to calculate the application costs, consider the costs and time to feed stored feed, and the utilization of the stockpiled forages and the stored feed. In many cases, stockpiling is a viable option to reduce costs and save time.

<u>Literature Cited</u>:

Landefeld, M. & Penrose, C. (2016). Comparing Fertilizer Additives Applied to Urea for Stockpiling Orchardgrass. Available at: http://www.nacaa.com/journal/post_editor.php?article_id=549

McInnes, K. J., R. B. Ferguson, D. E. Kissel, and E. T. Kanemasu. 1986. Field Measurements of Ammonia Loss from Surface Applications of Urea Solution to Bare Soil1. Agron. J. 78:192-196. doi:10.2134/agronj1986.00021962007800010038x

Penrose, C., Landefeld, M., McCutcheon, J. & Lima, D. (2018). Fertilizer Options for Stockpiling Cool Season Grass. Available at: corn.osu.edu

Penrose, C., McCutcheon, J. & Landefeld, M. (2015). Utilizing Urease Inhibitors with Urea to Stockpile Fescue. Available at: http://www.nacaa.com/journal/index.php?jid=549

Table 1. Three Site Average

Treatments	lbs DM/acre	lbs DM/acre above control	CP%	ADF%	TDN%
Control	2645	-	11.81	41.16	58.89
Urea	3322	677*	12.05	43.00	58.09
Urea+Agrotain®	3494	849*	12.01	41.25	58.57
Ammonium Sulfate	3278	633*	12.12	42.79	58.08

LSD = 566 *(P* < 0.05) * denotes significant difference in yield compared to the control.

SCREENING CHEMICAL CONTROL OPTIONS FOR THE BERMUDAGRASS STEM MAGGOT

Lisa Baxter, Holly Anderson, Andy Carter, Philip Edwards, Jeremy Kichler, Jenna Kicklighter, Carole Knight, Will Lovett, Peyton Sapp, Andrew Sawyer, and Dennis Hancock[1]

The recent invasion of the bermudagrass stem maggot (BSM; *Atherigona reversura* Villenueve) has severely damaged bermudagrass (*Cynodon dactylon*) pastures and hayfields. Strategically-timed pyrethroid applications can significantly reduce adult BSM populations, but overuse of a single mode of action can lead to resistance. Cooperating producers and Extension Agents/Coordinators from eight South GA counties screened potential alternatives insecticides. Zeta-cypermethrin, malathion, and carbaryl were tested in 2017 and 2018. Spinosad was only tested in 2017 before it was replaced with chlorantraniliprole in 2018. In this study, zeta-cypermethrin reduced the BSM population by 89% after two applications. While pyrethroids are still the most effective MOA, malathion could be a suitable alternative. The other three insecticides will not be recommended. Chlorantraniliprole may be effective if used in conjunction with a pyrethroid, but it does not provide residual control for the BSM. Corresponding samples for each spray application were collected from an unsprayed control area. Significant differences were not found because of the variability in these samples, but it is important to point out that the BSM population only decreased in the unsprayed control when chlorantraniliprole was sprayed.

[1]Post-doctorate, University of Georgia, Tifton, GA (baxterl@uga.edu); Ben Hill Cty. Ext. Agent, Fitzgerald, GA, (hhickey@uga.edu); Lowndes Cty. Ext. Agent, Valdosta, GA, (aec22711@uga.edu); Irwin Cty. Ext. Coordinator, Ocilla, GA, (pedwards@uga.edu); Colquitt Cty. Ext. Coordinator, Moultrie, GA, (jkickler@uga.edu); Colquitt Cty. Ext. Agent, Moultrie, GA, (jkickli2@uga.edu); Bulloch Cty. Ext. Agent, Statesboro, GA, (clh@uga.edu); Bacon Cty. Ext. Coordinator, Alma, GA, (welovett@uga.edu); Burke Cty. Ext. Coordinator, Waynesboro, GA, (psapp@uga.edu); Wilcox Cty. Ext. Coordinator, Rochelle, GA (agsawyer@uga.edu); Forage Ext. Specialist, University of Georgia, Athens, GA, (dhancock@uga.edu)

SETARIA PARVIFLORA CONTROL IN TALL FESCUE

W.B. Greer, L. Dillard, J.A. Tredaway[1]

Setaria parviflora, more commonly known as marsh bristlegrass or knotroot foxtail, is a problematic perennial weed in forage crops grown for hay. The objective of these studies was to evaluate different herbicides for their control or suppression of *Setaria parviflora* in tall fescue. This trial was established in 2017 and repeated in 2018. To evaluate the benefits of a two year program, the 2017 trial was repeated in the same area in 2018. A similar trial was initiated in 2018 with additional herbicides. Dormant applications of Prowl H_2O, and Esplanade were applied alone and in sequential programs (after first cutting) with Facet L, and Facet L + Prowl H_2O. Stand-alone applications of Armezon at 1 and 2 fl oz/A were applied after first cutting in 2018. In both years, Prowl H_2O and Esplanade provided early season suppression of knotroot foxtail. After 224 days (65 days after cutting), the dormant applications alone had lost acceptable head suppression (40% or less). However, when a postemergence application was added to these dormant applications, the seed head suppression was evaluated at 82.5% or above. The standalone applications of Armezon also showed good to excellent control (82-87.5%). Although there was significant injury observed early with both of these treatments, complete recovery from the bleaching injury was observed by day 224. If a Prowl H_2O + Facet L or Armezon program can be developed to suppress seed heads or completely control knotroot foxtail, it would be extremely beneficial for tall fescue producers in the South.

[1]Research Associate, Auburn University, 201 Funchess Hall, Auburn University, AL 36849 (wbg0007@auburn.edu); Assistant Professor & Extension Forage Specialist, Auburn University/ACES, Upchurch , Auburn University, AL 36849 (dillasa@auburn.edu); Ph.D., Weed Research Solutions, LLC., Jackonville, AL 36265 (tredaway@gmail.com).

SIMPLIFYING GRAIN DRILL CALIBRATION

J.M. Buckman, H. Adams, and C.D. Teutsch[1]

Grain drill calibration is a critical, yet often ignored part of successful forage establishment and pasture renovation. Planting lower seed rates than recommended can result in thin stands that are susceptible to weed encroachment. Planting more than the recommended seeding rate is undesirable due to the high seed cost of improved forage varieties. Most farmers just use the seeding chart that is already on their grain drill. As drills wear, and tires and cogs get replaced, actual seeding rates can vary significantly from seeding charts found on drills. We have designed a simple and straight forward calibration method that can be applied across a wide range of grain drill types and manufacturers. This method is centered on a pre-made chart that allows producers to determine the quantity of seed to catch for each disk opener for a desired seeding rate. This minimizes the need for producers to carry out detailed mathematical calculations. In order to calibrate the drill using this method, you will need the following items: a container to catch the seed; tape measurer to determine the circumference of the drive wheel and the disk opener spacing; flags to mark stopping and starting points for in field calibration; a floor or bottle jack for stationary calibration; and a gram scale with 0.1 gram accuracy. This procedure and chart were made into a decal that can be affixed to grain drills. This decal has been distributed to counties and Soil and Water Conservation districts in Kentucky that have drills that are loaned or rented. A copy of the procedure/chart and an informational video can be found UK Master Grazer Webpage.

[1]Undergraduate, Animal Sci., Murray State Univ., Murray, KY (jbuckman10@murraystate.edu); Undergraduate, Agricultural Systems Tech., Murray State Univ., Murray, KY; Associate Extension Professor, Dept. of Plant and Soil Sciences, Univ. of Kentucky, Lexington.

SOFT RUSH CONTROL IN PERENNIAL WARM-SEASON GRASS PASTURES

E.K. Twidwell, R.E. Strahan, M.D. Voitier and A.L. Granger[1]

Soft rush (*Juncus effuses*) is a clump-forming perennial plant that often infests low-lying areas in Louisiana pastures. This species is commonly referred to as bull rush. It grows to a height of 2 to 5 feet, and reproduces by seeds and rhizomes. It is not a palatable species for livestock to consume. Control measures include clipping and herbicide treatment. Clipping provides some level of control, but is not a long-term solution. The objective of this study was to compare 7 herbicide treatments for control of soft rush in a pasture composed primarily of carpetgrass (*Axonopus affinis*). Treatments were applied on July 24, 2018. The experimental design was a randomized complete block with three replications. Four of the treatments were broadcast-applied and three were spot-treated on individual soft rush plants. Visual control ratings were taken 36, 77 and 92 days after treatment. Spot-treating individual plants with 2% glyphosate provided near 100% soft rush control. Spot-treating with 2,4-D provided better control than using broadcast 2,4-D applications. The herbicides Permit and Outrider did not provide satisfactory soft rush control in this study. Results of this study suggest that soft rush can be effectively controlled via spot-treatment with either 2,4-D or glyphosate herbicides.

[1]Professor, LSU School of Plant, Environmental and Soil Sciences (SPESS), Baton Rouge, LA (etwidwell@agcenter.lsu.edu); Associate Professor, LSU SPESS, Baton Rouge, LA (rstrahan@agcenter.lsu.edu); Research Associate, LSU SPESS, Baton Rouge, LA (mvoitier@agcenter.lsu.edu); Extension Agent, Vermilion Parish, Abbeville, LA (agranger@agcenter.lsu.edu).

SUBSURFACE APPLICATION OF POULTRY LITTER AND MOISTURE EFFECTS ON RYE AND RYEGRASS BALEAGE QUALITY

C.C. Nieman[ab], D.H. Pote[b], K.P. Coffey[c], W.K. Coblentz[d], and P.R. Owens[b]

Subsurface poultry litter application results in more efficient nutrient utilization compared to surface application and allows for greater forage yields and forage nutritive value with reduced nitrogen (N) run-off. Combining forage ensiling and storage as baleage with subsurface poultry litter application could result in even greater nutrient utilization by harvesting forage at peak quality in early spring when dry harvest is impossible. This study was conducted to compare yield and nutritive value of spring harvested rye and ryegrass preserved as baleage, when fertilized with poultry litter applied either subsurface or surface. Boot stage rye was cut in the afternoon of April 9, 2018 and baled between 11:30 am and 2:30 pm on April 10, 2018. Bale moisture ranged from 52.4 to 64.6% and averaged 58.7%. Yields differed between the two treatments: 0.9 and 1.5 tons DM/acre for surface and subsurface application methods, respectively. Yields on surface treatment plots were likely affected by dry conditions in fall, which allowed for partial N volatilization, while subsurface plots may have retained more N until growing conditions improved. However, plant nutritive composition was not affected by treatment, as there were no differences between crude protein, neutral detergent fiber, or acid detergent fiber in the pre-ensiled samples. Although, post-ensiled sample analysis may show differences in fermentation variables with subsurface poultry litter application, subsurface poultry application greatly increased yield, but did not affect forage nutritive value pre-ensiling.

[a]Postdoctoral Researcher, Oak Ridge Institute for Science and Education, Oak Ridge, TN; [b]ARS USDA, Dale Bumpers Small Farms Research Center, Booneville AR; [c]University of Arkansas, Department of Animal Science, Fayetteville, AR; [d]ARS USDA US Dairy Forage Research Center, Marshfield, WI.

SUGAR CANE APHID PRESSURE AND FORAGE PRODUCTION AND QUALITY OF SUMMER ANNUAL MIXTURES

S. Thomspon[1], A. Jacobson[2], and L. Dillard[1]

Summer annual forages such as sorghum, sudangrass, and sorghum × sudangrass hybrids have recently been seeing reduced yields due to sugarcane aphid infestations, which is leading Alabama producers to seek alternative forages that are not susceptible to sugarcane aphid infestations. This study was conducted to determine the yield, nutritive value, and sugarcane aphid infestation rates of different summer annual mixtures. The forage treatments (n = 4) used in this study were sorghum × sudangrass (S), S + insecticide (SI), cowpea (C), crabgrass (R), S + cowpea (S+C), S + crabgrass (S+R), cowpea + crabgrass (C+R), and S + cowpea + crabgrass (S+C+R). Sugarcane aphid counts were conducted weekly on plots containing S. Forage was dried and analyzed for crude protein (CP), acid detergent fiber (ADF), and neutral detergent fiber (NDF). There was no difference ($P > 0.8$) in the forage yield of S, S+R, S+C, S+C+R, C+R, and C, with the yield of R ($P < 0.05$) being lower than those. The NDF concentration was greatest ($P < 0.0001$) in the treatments containing S compared to plots without (64.2 and 57.3 +/- 1.00%, respectively). Cowpea and C+R had the lowest ($P < 0.02$) NDF concentrations. The ADF concentrations were greatest in plots containing mono- or bi-cultures of S. Mean aphid-days per two leaf sample was greater ($P < 0.0001$) in S+C plots compared to S+R and SI on July 12. While mixtures containing S did not have reduced sugar cane aphid pressure, C+R showed promise as a summer annual mixture with similar yield and higher quality to that of S and S mixtures.

[1]Dept. of Animal Sciences, Auburn University (sjt0015@auburn.edu, dillasa@auburn.edu);
[2]Dept. of Entomology and Plant Pathology, Auburn University (alj0043@auburn.edu).

SYNERGISTIC RELATIONSHIPS BETWEEN GRAZING MANAGEMENT AND INSECT BEHAVIOR

K.H. Robertson[1]

Grazing management has a cascading symbiotic effect on insects which in turn positively affects our pastures. Three examples include 1) spiders, 2) dung beetles, and 3) ground beetles. Some of these interactions are simple and directly based on the needs of the insect, and some are more complex, needing an extra step in the process. 1) The case using spiders is riveting. With long rest times and plenty of forage, spider populations find hospitable habitat. As the spider population increases, it moves the grasshopper population further up in the sward canopy. This means that the grasshoppers change from dining on grasses to dining on goldenrod. Keeping that canopy from shading the understory is shown to improve plant diversity because the lower plants are allowed the sunlight they require to germinate and thrive. This diversity was in turn found to increase nitrogen supply to the soil! 2) Studies have shown that dung beetles are more prevalent with good grazing management. The beetles are able to find the manure easily, and also find each other easily (which in turn means more breeding). Conventional grazing does not afford these amenities to the beetles. 3) In a field planted with cover crops (read 'tall forage'), beetles and rodents eliminated 3-4 times as many weed seeds as in a field that wasn't cover cropped (read 'bare soil'). In addition, when the field has cover the beetles tended to eat 50% more seed. This is likely due to fear of rodents which made the beetles hunker onto a pile of seeds instead of foraging for better food. When mimicking nature with our grazing management, we are leveraging synergies of all kinds to make healthier forage with fewer weeds.

[1]Civil Engineer/Farmer, Greenville County Soil & Water Conservation District, Greenville, SC.

THE EFFECTS OF GIBERELLINS AND UREA ON SPRING FORAGE GROWTH

D.F. Lima, C.H. Gelley, M.A. Landefeld, J. S. McCutcheon[1]

Gibberellins are compounds classified as plant growth regulators. This class of compounds has been shown to regulate multiple plant developmental factors such as breaking seed dormancy, and stem and leaf elongation. Our experimental study focused on increasing spring forage growth utilizing RyzUp SmartGrass®, a gibberellin based product, and fertilizer as dual growth stimulants during the spring of 2017 and 2018. Data from 2017 suggested a synergistic effect of RyzUp and urea. To further investigate this phenomenon, a follow-up experiment to test a low (0.4 oz) verses a high dose (1.0 oz) of RyzUp was conducted. Four treatments were implemented; a control with no added urea or RyzUp, urea applied at 100 lb/A, RyzUp applied at a 0.4 oz/A, and both RyzUp and urea applied together in the early spring. A mirrored experiment with 1.0 oz of RyzUp was conducted concurrently. We hypothesized a synergistic effect utilizing RyzUp SmartGrass® and urea. When tested at the 0.4 oz/A, the application of RyzUp and urea together showed a significant increase ($P < 0.05$) in biomass vs. the other treatments. Regression analysis indicated that the synergistic effects of the growth regulator and fertilizer could be seen independent of environmental variances for separate different years ($r^2 = 0.98$). When tested at the 1.0 oz/A rate of RyzUp in 2018, the treatment interactions matched those observed in 2017 and 2018 at the 0.4 oz/A rate. Synergistic effect could be due to a dual stimulus of a hormone (RyzUp) and chemical (urea) acting independently on separate growth pathways.

[1]Ohio State University Educator, St. Clairsville, OH (lima.19@osu.edu); Ohio State University Educator, Caldwell, OH (gelley.2@osu.edu); Ohio State University Educator, Woodsfield, OH (landefeld.6@osu.edu); Ohio State University Extension Educator, Ohio State University Assistant Director of Operations, Caldwell, OH (mccutcheon.30@osu.edu).

TURFGRASSES FOR HIGH TRAFFIC AREAS ON EQUINE OPERATIONS

A.L. Jaqueth, T.R. Turner, M.E. Iwaniuk, B.J. McIntosh and A.O. Burk[1]

Horses are large, athletic animals with a propensity for damaging foliage due to heavy hoof traffic and frequent grazing. Additionally, man-made areas bare of vegetation referred to as "dry lots" are commonly used for equids under dietary restriction (e.g. total diet NSC <10-12%) who cannot safely graze pasture due to fluctuating levels of non-structural carbohydrates (NSC). A lack of vegetation has been associated with negative environmental impacts including soil erosion, which could be prevented by anchoring soil through vegetation. Turfgrasses, such as those used for athletic fields, may be suitable as ground cover in dry lots and areas of heavy hoof traffic as they tend to be tolerant of traffic and close mowing (e.g. grazing) and may be low to moderate in both yield and NSC. The objective of this study was to evaluate 8 cool-season (CS) and 6 warm-season (WS) cultivars exposed to three levels of simulated hoof traffic. Suitability for use on equine operations was assessed by persistence and biomass, which were used to determine relative traffic tolerance ranking and nutritional composition. Persistence was reduced in response to traffic in CS cultivars, but not in WS cultivars. Biomass of CS cultivars was reduced in response to traffic for most trials, but for WS cultivars was increased during the first year of the study. Overall, tall fescue (*Schedonorus arundinaceus* (Schreb.) Dumort.) and zoysiagrass (*Zoysia japonica*) cultivars were most traffic tolerant, but only zoysiagrass had a more favorable NSC concentration of <15%.

[1] Graduate Teaching Assistant, The University of Maryland, College Park, MD (ajaqueth@umn.edu); Associate Professor, The University of Maryland, College Park, MD (tturner@umd.edu); Graduate Teaching Assistant, The University of Maryland, College Park, MD (miwaniuk@umd.edu); Equine Extension Specialist, Virginia Polytechnic Institute and State University, MARE Center, Middleburg, VA (bmcintosh@vt.edu); Associate Professor, The University of Maryland, College Park, MD (amyburk@umd.edu).

UGA FORAGE EXTENSION TEAM INCREASES PROGRAMING OPPORTUNITIES BY USING A UNIQUE STRUCTURE AND INNOVATIVE PROGRAMING TECHNIQUES

Morgan, S.[1], Speir, A.[2], Hicks, R.[3], Knight, C.[4], Ray, L.[5], Kichler, J.[6], Gates R.[7], Meeks, C.[8], Hancock, D.[9]

Forage crops are important to the state of Georgia. There are approximately 4 million acres of pasture, hay and silage in Georgia (10% of the state). The UGA Forage Extension Team works to assist in providing unbiased, research-based information to forage producers across the state in a timely and effective manner. Team members, strategically located throughout the state, are a specialized group of Extension Agents selected based on their knowledge of forages. The team conducts forage programming in their respective areas or regions of the state. The UGA Forage Extension Program was assessed through an online survey during February and early March of 2017. Participants self-reported that their participation in UGA Forage Extension programs were directly responsible for increasing their net farm incomes by an average of 16.1% (+/- 2.3) and net agribusiness income by 17.1% (+/- 2.3). Since 2015, the UGA Forage Extension Team has made 2773 face to face contacts. The GrassMasters program has reached 128 producers face to face. Through the quarterly newsletter, team members are making a regional impact by providing educational information to over 7200 forage producers and industry professionals.

[1]Harris County Extension Coordinator, UGA Cooperative Extension, Hamilton, GA 31811;
[2]Madison County Extension Coordinator, UGA Cooperative Extension, Danielsville, GA 30633;
[3]Screven County Extension Coordinator, UGA Cooperative Extension, Sylvania, GA 30467;
[4]Bulloch County Extension Agent, UGA Cooperative Extension, Statesboro, GA 30458;
[5]Morgan County Extension Coordinator, UGA Cooperative Extension, Madison, GA 30650;
[6]Colquitt County Extension Coordinator, UGA Cooperative Extension, Moultrie, GA 31788;
[7]Whitfield County Extension Agent, UGA Cooperative Extension, Dalton, GA 30720;
[8]Houston County Extension Coordinator, UGA Cooperative Extension, Perry, GA 31069;
[9]Extension Forage Specialist, UGA Cooperative Extension, Athens, GA 30602.

UGA GRASSMASTERS: AN EDUCATIONAL PARTNERSHIP TO DEVELOP BETTER GRAZING MANAGEMENT IN GEORGIA

Knight, C.H.[1], Brown, P.[2], Gates, R.[3], Hancock, D.[4], Hicks, R.[5], Kichler, J.[6], Meeks, C.[7], Morgan, S.[8], Ray, L.[9], and Speir, A.[10]

Forage production is a big part of the farm gate of Georgia, representing approximately $130 million and 4 million acres. The UGA Forage Extension Team, comprised of state county agents, works to meet the diverse demands of clientele in Georgia. In 2016, the UGA Forage Extension Team collaborated with USDA-NRCS and the Two Rivers Resource Conservation and Development Council to present a new program to complement existing UGA forage programs. The UGA GrassMasters Program serves as an in-depth introduction to forage production. The seven-week series serves as an introduction to more advanced courses, like Grazing School. The UGA Forage Extension Team created the curriculum and teach the training modules with the NRCS State Grasslands Conservationist. Since its inception, five GrassMasters series have been held reaching 142 participants. A measure of knowledge gained was evaluated using a Likert-type scale from 5 to 1. A score of 5 indicated that knowledge was greatly improved and a score of 1 indicated no change in knowledge. All topics received a score of 4 or higher. Follow-up surveys indicate that a majority of participants have made changes to their operations and consider the program a major help to attaining their forage management goals.

[1]Ag and Natural Resources Agent, University of Georgia (UGA) Extension, Bulloch County, Statesboro, GA 30458; [2] Grasslands Conservationist, Georgia NRCS, Athens, GA 30602; [3] Ag and Natural Resources Agent, UGA Extension, Whitfield County, Dalton, GA 30720; [4] Extension Forage Specialist, UGA Crop and Soil Science, Athens, GA 30602; [5] Ag and Natural Resources Agent, UGA Extension, Screven County, Sylvania, GA 30467; [6] Ag and Natural Resources Agent, UGA Extension, Colquitt County, Moultrie, GA 31788; [7] Ag and Natural Resources Agent, UGA Extension, Houston County, Perry, GA 31069; [8] Ag and Natural Resources Agent, UGA Extension, Harris County, Hamilton, GA 31811; [9] Ag and Natural Resources Agent, UGA Extension, Morgan County, Madison, GA 30650; [10] Ag and Natural Resources Agent, UGA Extension, Madison County, Danielsville, GA 30633

USING A MILLET NURSE CROP FOR NATIVE GRASS ESTABLISHMENT

J.D. Richwine[1], P. Keyser[2], D. Hancock[3], and L. Marks[4]

Incorporating a nurse crop with native warm-season grasses (NWSG) may suppress weeds and allow forage production to continue during the seedling year. Therefore, we conducted two NWSG studies at East Tennessee AgResearch and Education Center in Knoxville, TN, from 2016-2017, and replicated 2017-2018, to evaluate big bluestem (BB) [*Andropogon gerardii* Vitman] and switchgrass (SG) [*Panicum virgatum* L.] establishment when a browntop millet (BTM) [*Urochloa ramosa* (L.) Nguyen] nurse crop is incorporated. Each study was a randomized complete block with a 2x3 factorial arrangement of treatments. Treatments included two canopy management strategies (harvests based on BTM maturity and competition with NWSG seedlings) and three BTM seeding rates (0, 10, and 20 lbs/ac). Native WSG seedlings/ft^2 were counted 30 and 60 days post-planting and following dormancy. Only BTM seeding rate was significant among treatments. At 60 days post-planting, both BB (P = 0.0065) and SG (P = 0.0434) showed differences among BTM seeding rates while only BB (P = 0.0398) was different following dormancy. In all cases, 0 lbs/ac had more NWSG seedlings/ft^2 than 20 lbs/ac. These results indicate that planting a BTM nurse crop at 20 lbs/ac would impede BB establishment more than not planting BTM.

[1]Graduate Research Assistant, Center for Native Grasslands Management, Dept. of Forestry, Wildlife, and Fisheries, University of Tennessee – Knoxville, Knoxville, TN (jondrich@vols.utk.edu); [2]Professor, Center for Native Grasslands Management, Dept. of Forestry, Wildlife, and Fisheries, University of Tennessee – Knoxville, Knoxville, TN (pkeyser@utk.edu). [3]Associate Professor, Dept. of Crop and Soil Sciences, University of Georgia, Athens, GA (dhancock@uga.edu); [4]Regional Extension Agent, Alabama Cooperative Extension System, Centre, AL (mlm0013@aces.edu)

USING SUMMER ANNUALS TO TRANSFORM FORAGE SYSTEMS IN WESTERN KENTUCKY

H. Adams, J.M. Buckman, and C.D. Teutsch[1]

Forage systems in the transition area between the temperate north and subtropical southern U.S. are based on cool-season grasses that are productive grazing during the spring and fall, but have limited growth during the summer months. In contrast, warm-season annual grasses and legumes are highly productive during the summer months. The objectives of this project were to introduce cattle producers to improved summer annual cultivars and to demonstrate how they could fit into forage systems in Western Kentucky. Working with local extension agents, the demonstrations were conducted on five farms in Western Kentucky. In late May and early June, a BMR sudangrass (var. 'AS9302'), a pearl millet (var. 'Wonderleaf'), a forage soybean (var. 'Large Lad'), and a mixture of three were planted in 2-acre strips on each of the five farms. When the plots reached 30 to 40 inches in height, they were sampled for yield and nutritive value and grazing was initiated. The demonstration areas were subdivided with temporary fencing and rotationally stocked. Averaged over locations, sudangrass, pearl millet, forage soybean, and the summer annual mixture yielded 5,138, 5,259, 2,234, and 4,654 lb DM/A, respectively. At three of the five locations, a summer field day was held to highlight the use warm-season annuals in grazing systems. Data will be presented this winter at local extension meetings.

[1]Hunter Adams, Agricultural Systems Technology, Murray State University, Murray, KY (hadams@murraystate.edu); Undergraduate Student, Animal Science, Murray State University, Murray, KY (jbuckman@murraystate.edu); Associate Extension Professor, Dept. of Plant and Soil Sciences, University of Kentucky, Lexington (chris.teutsch@uky.edu).

USING THE KENTUCKY FORAGE COUNCIL BOARD TO SET UK EXTENSION PROGRAMMING AND AGENT TRAINING PRIORITIES

Jimmy C. Henning, Ray Smith, Chris Teutsch and Traci Missun[1]

The Kentucky Forage and Grassland Council (KFGC) has been providing agricultural leadership to Kentucky since 1964. In the fall of 2016, KFGC was asked to provide programming input to the University of Kentucky College of Agriculture Food and Environment as preparation for the development of the federally-required Extension Plan of Work. The forage council surveyed members online via Qualtrics®. An ad hoc committee of KFGC leadership consolidated the full survey results into five major themes. UK specialists led the KFGC board through a 'sticky note' exercise to prioritize these themes and to recommend specific programming needed. An unintended benefit of the advisory exercise was identification of high priority training needs for county extension agents. The top five areas of forage programming as ranked by the KFGC board of directors were: 1) Alternative forages and grazing cover crops, 2) Economics of forage production practices, 3) Silage, haylage and baleage and pricing for stored feed, 4) Weed identification and herbicide information, and 5) Using novel endophyte tall fescue. UK specialists report on progress towards these objectives at each board meeting.

[1]Extension Professor, University of Kentucky, Lexington, KY (jimmy.henning@uky.edu); Extension Professor, University of Kentucky, Lexington, KY (raysmith1@uky.edu); Extension Associate Professor, University of Kentucky Research and Education Center, Princeton, KY (chris.teutsch@uky.edu); University of Kentucky County Extension Agent for Agriculture and Natural Resources, Oldham County, KY (traci.missun@uky.edu).

WE NEED A RUMINANT REVOLUTION!

P.J. Ballerstedt[1]

Human beings exist because of ruminants. Today's societies rely upon them. Humanity's future depends upon improvements in the productivity and efficiency of worldwide ruminant animal agriculture. Like the general public's confusion of what constitutes a "healthy diet," tremendous misunderstanding exists regarding the environmental role of ruminant animals. Diets rich in butter, meat and cheese have been shown to promote human health and development. Such diets can correct the symptoms of metabolic syndrome, offering hope in arresting the current worldwide epidemic of chronic diseases. The Dietary Guidelines for Americans, however, are based on the belief that plant-based, high-carbohydrate diets are "healthy." While nutrition research does not support the belief that vegetarian diets are healthier than animal-based diets, an appeal to "sustainability" is now used to justify these recommendations. Sustainability must include societal, economic, and ecological aspects, including the enormous suffering and financial costs of chronic illness. We need a revolution in our thinking of what constitutes a healthy diet, of the vital role that animal products play in the human diet, and the essential nature of ruminant animal agriculture in meeting humanity's needs. This will mean overthrowing established policies and institutions. We need an effort, analogous to the Green Revolution, to develop and deploy the knowledge and technology necessary to meet the mid-21st Century demand for ruminant animal products while preserving and enhancing the environments in which they're produced.

[1]Forage Ambassador, Barenbrug USA, PO Box 233, HWY 99E, Tangent, OR (pballerstedt@barusa.com).

WRAP-TYPE AFFECTS DRY MATTER AND NUTRITIVE VALUES OF ALFALFA HAY IN LONG-TERM OUTDOOR STORAGE

A.S. Reiter, M.S. Wells, C.C. Sheaffer, A.M. Grev, K.L. Martinson[1]

It has been well established that storage conditions and bale wrap-type influence dry matter (DM) and forage quality in round bales stored outdoors. However, recent advances in bale wrap technology and plant genetics warrant new investigation. B-Wrap® is a microscopic pore material covered with net wrap that is marketed to shed precipitation without trapping moisture in the bale. Additionally, reduced lignin alfalfa varieties are being marketed to increase forage digestibility. Though, these technologies have yet to be evaluated in round bales stored outdoors long-term. The objective of this research was to evaluate changes in DM and forage nutritive values of conventional and reduced lignin alfalfa round bales wrapped in plastic twine, net wrap, and B-Wrap® under long-term, outdoor storage. The two alfalfa varieties yielded 24 round bales, resulting in four bales of each wrap-type for both alfalfa varieties. During harvest, the time to wrap a bale in each wrap-type was recorded. To estimate DM and forage nutritive values, bale weights were recorded and hay cores were taken to depths of 0 to 6 inches and 7 to 18 inches every 3 months for 12 months. All bales were stored outdoors on pallets. The twine wrap required the most time to tie a bale compared to B-Wrap®, while net wrap required the least amount of time (56, 28 and 18 seconds, respectively). After 12 months of storage, the outer 6 inches of B-Wrap® bales maintained DM percentage and nonstructural carbohydrates (NSC) compared to a reduction in both twine and net wrap bales. Interestingly, no differences in percent DM and forage nutritive values were observed at bale depths 7 to 18 inches. These results indicate that B-Wrap® has the potential to maintain DM and forage nutritive value in the outer 6 inches of alfalfa round bales in long-term outdoor storage.

[1]Graduate Research Assistant, Department of Animal Science, (reit0162@umn.edu); Associate Professor and Professor, Department of Agronomy and Plant Genetics, (mswells@umn.edu; sheaf001@umn.edu); Graduate Research Assistant, Department of Animal Science, (grevx004@umn.edu); Professor, Department of Animal Science, (martinson@umn.edu) University of Minnesota, St. Paul, MN, USA

YIELD, NUTRITIVE VALUE, AND NITROGEN REPLACEMENT VALUE OF ANNUAL LEGUMES

K.A. Cassida[1]

We evaluated annual forage legumes for two vital roles in Michigan agriculture: feed for livestock and green manures to provide nitrogen for subsequent crop rotations. Seven annual or perennial legumes were planted in spring and late summer for two years at three sites in Michigan: south central (East Lansing, EL), north central (Lake City, LC), and Upper Peninsula (Chatham, CH). An annual ryegrass hay crop was grown the year after legumes in order to assess nitrogen fertilizer replacement value. Legume establishment was adequate for cover crops for all plantings, but late planting did not consistently produce harvestable forage in CH. Spring planting produced more harvested forage than summer planting at all sites, but spring plantings contained more weeds. On a pure legume basis, yields did not differ between planting dates at EL or LC. Regrowth allowed two cuttings from spring plantings in EL, but was not dependable at more northern sites. Among sites, berseem clover had greatest pure clover yields at EL (up to 1.47 ton/acre), while crimson clover performed best at LC and CH (up to 1.74 ton/acre). Red clover yielded up to 0.79 ton/acre at EL and LC. Arrowleaf and balansa clovers did not have dependable yield as forage at any site. Summer legume seeding provided more fertilizer nitrogen replacement value for a following grass hay crop (up to 130 lb/acre in 2017) than spring seeding for berseem, crimson, arrowleaf and balansa clovers. Sainfoin was not acceptable for either purpose under these growing conditions.

[1]Forage and Cover Crop Extension Specialist, Michigan State University, East Lansing, MI (cassida@msu.edu).

YIELDS OF OVERSEEDING LEGUMES X FORAGE BERMUDAGRASS COMBINATIONS

W.F Anderson and F. Cheek[1]

Quality of forage for grazing or hay can be improved by having mixtures of grass with legumes. Clovers and alfalfa can be over-seeded into perennial bermudagrass in many parts of the South. The objective of this test was to test crimson clover and two alfalfa varieties developed for the Southeast in combination with three forage bermudagrasses for yield. Russell, Coastal and Tifton 85 bermudagrass plots were over-seeded with forage legumes (crimson clover and Bulldog 505 and Bulldog 805 alfalfa) in a randomized complete block design with four replications for a period of three years. One control treatment of bermudagrass only with inorganic N, P, and K was included for each bermudagrass entry. Legumes were over-seeded into the stubble in October. Plots were harvested and sampled five times each year in 2015 and 2016. A foot square block was harvested by hand to determine the percentage of grass and legume by separating the species by hand and processing the samples by taking wet and dry weights. Legume establishment was limited in 2015 with only 10 to 15% of the total yield being from the alfalfa and crimson clover. In 2015 Tifton 85, Russell and Coastal with inorganic fertilizer produced 26,000, 21,000 and 18,000 lb/acre respectively, and over-seeding treatments averaged 8,850 and 8,900 and 7,000 lb/acre, respectively. In 2016 the clover plots produced high yields in the first two harvests. Grass fertilized with inorganic N produced the highest yields (Tifton 85 - 12,000 lb/acre; Russell – 11,000 lb/acre and Coastal 8,600 lb/acre), the higher persistence of alfalfa (19 to 50% of total) and clover (44 – 75% of total) in 2016 produced between 4,400 and 7,200 lb/acre of high quality forage.

[1]Research Geneticist (bill.anderson@ars.usda.gov) and Research Technician (Freddie.cheek@ars.usda.gov), USDA-ARS, Tifton, GA

YOUNG CATTLEMAN SERIES OF CARROLL COUNTY

O. Foster[1]

A need arose for a program to assist younger producers in learning best management practices for their beef cattle operations. A base curriculum was developed that targeted the following areas: finances, resources (FSA, NRCS, UA, etc.,) herd management, herd improvement, forage management, pest management, and marketing. Carroll County met once a month from November to February for six hour sessions and limited the class size to fifteen. There were farm visits conducted with all participants and an end of course survey. At the conclusion of the program in Carroll County, the end of course survey and exit interviews were conducted with the following results: 100% of participates plan to implement a new practice learned from the class that wasn't present in their operation before, 100% of participants plan to utilize extension resources when moving forward. One producer even stated he changed his feed rations at the beginning of December after a class and has already saved several hundred dollars in feed cost at the conclusion of the program. According to survey results, extension educators played an important role in demonstrating the benefits and need of best management practices in all aspects of an operation.

[1]County Extension Agent - Agriculture, University of Arkansas, Cooperative Extension Service, Berryville, AR 72616.

AUTHOR INDEX

AUTHOR INDEX